The Debates Shaping
Spectrum Policy

The Debates Shaping Spectrum Policy

Edited by Martin Sims

CRC Press
Taylor & Francis Group
Boca Raton London

CRC Press is an imprint of the
Taylor & Francis Group, an **informa** business

First edition published 2022
by CRC Press
6000 Broken Sound Parkway NW, Suite 300, Boca Raton, FL 33487–2742

and by CRC Press
2 Park Square, Milton Park, Abingdon, Oxon, OX14 4RN

© 2022 Taylor & Francis Group, LLC

© 2022 selection and editorial matter, Martin Sims; individual chapters, the contributors

CRC Press is an imprint of Taylor & Francis Group, LLC

Library of Congress Cataloging-in-Publication Data

A catalog record has been requested for this book.

ISBN: 9780367742478 (hbk)
ISBN: 9780367742508 (pbk)
ISBN: 9781003156765 (ebk)

DOI: 10.1201/9781003156765

Typeset in Adobe Caslon Pro
by Apex CoVantage, LLC

Cover illustration: Tom Bull

For Jean and Ron

Contents

Preface

After completing *Understanding Spectrum Liberalisation* (CRC Press, 2015) with Toby Youell and Richard Womersley, I had one of those "Actually, what you should have done . . . " moments. The book gave a history of the introduction of spectrum liberalisation and tried to assess where it had succeeded and failed.

My concern was whether people would still want to read it in five years' time. This was a book, not the news articles and research reports about spectrum policy produced in my day job at *PolicyTracker*. These are designed for consumption within weeks or months, but books need to have a value that lasts for years.

It probably wasn't the fairest of criticisms, as we are always trying to learn lessons from the past, but it was the genesis of this new book. Could we identify the trends and tensions which will drive the evolution of spectrum policy over the next decade? Our contributors have kindly applied their many years of experience to this nigh-impossible task. Human beings can predict everything except the future, as the saying goes, and ironically Richard Womersley addresses this very issue in Chapter 5!

Both in researching this book and in our work at *PolicyTracker*, there is a group of people to whom I must extend both thanks and admiration. They are the people who agree to talk to us, often at no personal or commercial benefit, but because they recognise the importance of spectrum policy and are committed to an open and wide-ranging exchange of ideas.

The contributors to this volume are leading members of that group and I would like to thank them for giving so generously of their time. My colleagues at *PolicyTracker*, Manuel R. Marti, Toby Youell, Catherine Viola, Dianne Northfield, Richard Handford and Dugie Standeford, have worked tirelessly to research spectrum policy around the world and that – along with their insightful analysis – provides the basis for this book. I would also like to thank Tomoko Yamajo of KDDI for her continuing support of our research work.

Finally, a thanks to my family: Madhu, Chetan, Maya and Asha, without whom none of this would be possible.

And if you are reading this in 2026 or beyond, a particular thanks to you, as my dream of relative longevity will have succeeded.

Martin Sims
London
2021

Contributors

Martin Cave is a visiting professor in the Law Department of the London School of Economics and chair of the UK energy regulator, Ofgem. He has undertaken spectrum policy reviews and advised on spectrum auctions in a number of countries.

Dr Mohamed El-Moghazi is an acting executive director of national spectrum affairs at NTRA of Egypt. He is currently vice-chairman of the CPM-23, chair of ASMG Working Group 1 focusing on broadband and mobile issues and chair of ATU Task Group on spectrum recommendations.

Simon Forge is the managing director of SCF Associates Ltd that focuses on projects in telecommunications, computing and software, with their socio-economic and regulatory impacts, exploring new economic, radio and computing models and technologies, usually having an emphasis on planning and forecasting and often using the SCF scenario-based approach.

Dr Mary Longhurst is the managing director of Epoch Strategic Communications, which specialises in strategic communications and corporate social responsibility consultancy.

Manuel R. Marti is a journalist for *PolicyTracker* and has an MSc in International Politics from the School of Oriental and African Studies in London.

Marja Matinmikko-Blue is a 6G Flagship Research Coordinator and a senior research fellow at CWC, University of Oulu, where she holds an adjunct professor position on spectrum management. She conducts multi-disciplinary research on technical, business and regulatory aspects of mobile communication systems in close collaboration between industry, academia and regulators.

Gérard Pogorel is a professor emeritus of economics, Telecom Paris Graduate School of Engineering, CNRS Interdisciplinary Institute for Innovation. He co-founded the European Spectrum Management Conferences and acted as its chair/rapporteur. He recently published *A European Audiovisual Area for the World of Global Entertainment.*

Nicky Preston is a senior communications lead at Vodafone New Zealand and is responsible for the company's approach to stakeholder concerns around EMF.

Martin Sims is the managing director of *PolicyTracker*, which provides news, training and research about spectrum management.

Prof. William Webb is the CEO of Webb Search providing independent telecoms consultancy and advice. He was one of the founding directors of Neul and became CEO of the Weightless SIG. Prior to this, he was a director at Ofcom. He was IET president 2014–2015, has published 17 books and has multiple honorary doctorates and awards.

Dr Jason Whalley is a professor of digital economy at Northumbria University, Newcastle, UK. He is the editor of Digital Policy,

Regulation and Governance and vice-chairman of the International Telecommunications Society.

Richard Womersley is the managing director of LS Telcom UK and has been providing consultancy advice and training to regulators on the topic of spectrum management for over 20 years.

Toby Youell is a researcher and journalist on radio spectrum policy for *PolicyTracker*. A journalist by training, he has written extensively about spectrum since 2013 and also worked for the UK regulator, Ofcom.

INTRODUCTION

MARTIN SIMS

Contents

This book examines the debates that shape spectrum policy through a series of chapters from a range of notable experts. They consider not just how these issues *have* shaped policy but also how they are likely to continue to do so in the future, mapping out the motivations and challenges.

The contributors come from a variety of backgrounds, from industry to academia and from consultancy to journalism, enabling an assessment of the issues from a range of institutional perspectives.

This introduction identifies the debates and explains how the chapters address the issues. In selecting the topics, we have been deliberately wide-ranging. Engineering is vital to spectrum management but law, economics, public policy and strategic communications are also

DOI: 10.1201/9781003156765-1

1

important, and we have tried to reflect this, highlighting some areas that have not been widely discussed.

Are Spectrum Auctions Holding Back Infrastructure Investment?

Over the past 30 years, auctioning mobile spectrum has moved from heresy to orthodoxy, to paraphrase a famous article,[1] but from the mid-2010s that consensus has started to fracture. These doubts have their roots in the growing need for investment in telecoms networks, the seeds of which were sown in Europe as the continent fell behind the USA in 4G deployment. Concern has grown as we moved into the 5G era, where the use of higher frequencies means network densification with associated higher costs. Would high auction prices remove the cash which is needed for investment?

The backdrop to this fear is the growing social importance of mobile services. They are no longer the new boys throwing a free-market lifeline to a monopolistic wireline market with their promise of infrastructure competition. They are no longer a nice-to-have add-on to modern life, but the backbone of the future economy where an enormous range of everyday devices and everyday services – from healthcare to transport – will rely on ubiquitous high-speed connectivity.

Some of the required spectrum – particularly in mmWave – has still to be released in many parts of the world and hundreds of existing mobile licences will come up for renewal over the coming decades. Given the unprecedented need for mobile infrastructure investment, should these licences be auctioned?

The first two chapters in this book approach that question from different perspectives. Professor Martin Cave asks us to appreciate the economic benefits of spectrum auctions, to recognise what they have achieved not just in ensuring that the spectrum scarcity rent goes to governments to spend on socially beneficial causes but also in encouraging competition and widening access through the use of spectrum caps and coverage obligations.

Professor Gérard Pogorel takes the opposite starting point, accepting auctions' achievements in the early years of mobile but urging us to recognise their deficiencies in delivering the policy objectives which matter for the future, namely nationwide high-speed coverage. "The high-level, often repeated, argument that auctions are an effective process to select those operators that can use the frequencies most valuably, is as hard to verify as it is to falsify," he says.

Pogorel is persuaded by studies which suggest auctions can impair investment and fears that they have become a way of erecting entry barriers rather than encouraging competition. He is concerned that financial benefits for governments have obscured more important policy goals and argues that auctions should only be one element in a policy mix which priorities investment and coverage.

Interestingly, Cave's conclusions focus on ways to ensure that auctions maximise competition and deliver on social policy objectives, such as not restricting supply, including coverage and other obligations and using smaller lots. He also wonders how long auctions can survive if 5G requires ever more sharing and likely fewer operators.

Cave's and Pogorel's journeys start at opposite points but their final destinations seem very close!

How Can Spectrum Policy Encourage Innovation?

Encouraging innovation has become a widely accepted goal of spectrum policy: regulators want to encourage it and so does industry. Incumbents, start-ups and new entrants all claim to have the answers to the innovation conundrum but as their interests are antithetical can they all be right? "Innovation" has become so liberally sprinkled over policy documents that one wonders whether there is any consensus on what it means. Specifically, which regulatory policies encourage innovation and which hold it back?

Professor William Webb says this is a difficult question to answer, because "blocked innovations do not become visible and so demonstrate what could have happened had the regulator been more open to innovation." He argues that regulators are inherently risk averse, which is no bad thing in some circumstances, such as safety of life services, but tends to work against any innovation which requires major changes. Webb says regulators must recognise these conservative tendencies and try to change their cultures.

Do International Spectrum Policy Arrangements Need to Be Updated?

Spectrum policy is necessarily an international issue – radio waves do not respect national borders – and for over one hundred years the International Telecommunication Union (ITU) has been the main forum for global cooperation. But in that time the huge technological developments in wireless technology have not been matched by major changes in the international coordination of spectrum management. It remains focused on interference protection, harmonisation, consensus and spectrum identification, according to Dr Mohamed El-Moghazi and Professor Jason Whalley. These "generate benefits but can also be used to deter innovation and enforce 'traditional' ways of managing spectrum," the authors argue in their chapter "Towards a Future-Proof International Spectrum Policy?"

This considers whether the ITU procedures are too slow moving in an era when the pace of technology change is so fast, resulting in some countries making independent spectrum identifications outside of the World Radiocommunications Conference (WRC) process, as the USA famously did with 5G in 28 GHz.

The authors examine the tension between harmonisation and flexibility, where the drive to create economies of scale by coordinating the use of bands means some countries are denied the flexibility which would enable them to use frequencies in a way that best suits their national needs. An example of this is India, which was unable to make an International Mobile Telecommunications (IMT) identification in 600 MHz, despite reassurances that it would not cause interference to neighbours' broadcasting services. This calls for reviewing the ITU decision-making procedures especially the consensus concept, as argued by El-Moghazi and Whalley.

Another theme for the authors is whether current international practices are restricting innovation. They assess the process of identifying spectrum for some technologies and not others, wondering whether this may restrict new services. The static nature of the Radio Regulations and the relative lack of sharing is another concern. "Service allocations are fixed in large bandwidths for a period of decades," the authors say, because it is too difficult to change them. They suggest ways of circumventing this tendency as well as considering options for a wider review of ITU processes.

How Should We Use Demand Forecasting in Spectrum Policy?

When ITU members meet at WRCs, a key task is to allocate spectrum for services which hope to use it in the future. These conferences take place about every four years and any new spectrum proposal has to be agreed at the previous WRC. Furthermore, standards bodies may take several years to agree on the technological approach, so "the future" may be up to ten years away. Estimating the likely usage of any service that far in advance is extremely difficult and on occasions has been very inaccurate, as Richard Womersley points out in his chapter.

The most attention-grabbing example was the spectrum demand forecasts prior to WRC-19. The ITU had predicted that IMT would need up to 1960 MHz of spectrum and by 2019 had identified around 1300 MHz. However, on average, each country had typically licensed only 500 MHz of this and operators' networks seemed to be coping well, despite data growth exceeding many forecasts. Similar problems exist with models developed by various other organisations.

Womersley discusses where these errors came from, not just in a modelling sense but also in the organisational structure of the bodies which produced them. Over allocation is as much of a problem as under allocation, he says, as it means the inconvenience of re-farming for existing users and the possibility that this newly vacated band may never be used.

Womersley argues that demand forecasting is a necessary evil but warns that it must be treated with great caution.

Where Does Spectrum Fit into the Politics of the Global Economy?

Spectrum "is the arbiter of who can enter and compete in markets that contain some of the most profitable sectors on the planet," says Simon Forge, making it extremely important to governments and the global economy. His chapter talks about what this means in realpolitik terms.

Vendors get strong support from national industrial policies because of their extraordinary value, Forge argues: Apple was judged the world's most valuable company in 2020.[2] The cost of building a network and the high fees paid in spectrum auctions mean many mobile operators have come to depend on finance provided by equipment suppliers. Once 4G equipment was sold, Forge says, the race was on to generate new revenues by marketing 5G, with heavy backing from governments.

"From a *realpolitik perspective*," says Forge, "the 5G 'race' has been seeded by conditions in the global telecoms equipment industry emphasising the race is a supply side push rather than a demand side pull and fuelled by industrial policy among the major supplier nations."

But among mobile network operators (MNOs) and investors, there is scepticism about the returns which 5G can bring, manifesting itself in a lack of enthusiasm for auctioned mmWave 5G spectrum and a move towards using C-Band and below. Worry about capital expenditure (CAPEX) is also driving the lowering of costs through forms of spectrum sharing.

Forge concludes by asking whether 5G will meet future needs. Can it connect the unconnected billions – where Low Earth Orbit (LEO) satellites seem a better solution? And can it meet the need for improved network security exposed in recent cyberattacks – or will this have to wait for 6G?

How Will the Next Generation of Mobile Technology Affect Spectrum Policy?

At the time of writing, 5G had limited deployment in most developed countries, and the development of 6G was at very early stage. However, among the researchers working on 6G a consistent theme had emerged: the need to align 6G – due to be deployed in 2030 – with the UN Sustainable Development Goals (SDGs) and due to be achieved in the same year. The goals cover topics, such as poverty, hunger, gender equality, quality education and climate change.

In her 6G chapter, Dr Marja Matinmikko-Blue argues that 6G can help attain the SDGs not just through improving communications but also through combining these capabilities with sensing, location, imaging and other capabilities to report on the achievement of the goals.

What are the implications for spectrum policy? Matinmikko-Blue explains the necessity to improved stakeholder engagement, considering that the SDGs have objectives that go far beyond the interest of the telecoms industry and into wider social and environmental policy. She argues that a greater diversity of spectrum access regimes will be needed – particularly spectrum sharing – if these wider goals are to be met and new groups of vertical users accommodated.

The benefits of this wider stakeholder engagement are already apparent, she says:

> It can be observed that decisions made regarding 5G spectrum awards were dominated by the needs of those with existing strong market positions. Incumbent's views on future needs were considered above all others. In instances where input from different stakeholders was considered, and where innovation, experimentation and disruption were encouraged, new practices (e.g. local licensing and new obligations that benefit consumers being included as part of the auction process) created meaningful positive change.

Will Artificial Intelligence Change Spectrum Management?

Increasing the use of artificial intelligence (AI) is one of the key research areas for 6G, and Google's Loon project offered a fascinating vision of how it might be used in spectrum management in the future. Loon's software platform combined weather data, signal propagation models, hardware/antenna configurations, spectrum regulations and licensing information to enable the provision of broadband services without causing interference to planes and terrestrial users.

The use of AI in spectrum management is attracting funding from the US government and elsewhere and, in his chapter, Toby Youell charts its progress so far, assesses its potential and asks to what extent it can replace human activity.

His particular concern is unauthorised use of the airwaves, something which is currently limited by the high costs of developing equipment for new bands. The combination of AI and software defined radio may change that, Youell argues, and raise an upcoming enforcement problem which may limit the initial use of AI to specific locations and frequencies.

How Can Spectrum Policy Become Greener?

Spectrum policy has long had social goals such as reducing the digital divide but the idea that it might be able to benefit the environment is a recent one. Policymakers are increasingly recognising its importance but so far concrete measures have been limited, as Manuel R. Marti explains.

There is, however, an increasing recognition of the dual role played by information and communications technologies (ICT). They both create greenhouse gas emissions – principally through electricity consumption – and improve efficiency in other industries so reducing their environmental impact. 5G will certainly make mobile more energy efficient but there is a further paradox: while each mobile generation has reduced the energy per byte, the total amount consumed has gone up as subscriber numbers have increased and data usage soared.

How can spectrum policy help? Marti highlights some of the options being floated, including making energy efficiency a licence condition like coverage obligations. Spectrum policy must also play its part in controlling electronic waste, as extending the life of all smartphones in the EU by one year would be the equivalent[3] of taking a million cars off the roads in terms of CO_2 emissions. As phones come to operate in an ever-wider range of bands, should not a licence condition for using these new bands be improved recycling of old phones, Marti asks.

How Should We Inform the Public about Spectrum and Health Issues?

Two chapters examine the issue of public health messaging about spectrum. The spectrum management community has devoted considerable time and resources to ensuring that the use of mobile phones and other devices is not injurious to health, but a substantial section of the public has not been convinced. Public fears about the health risks have ebbed and flowed over the past two decades, with some arguing that the fiercest storms have roughly coincided with the launch of each mobile generation.

However, the wave of anti-5G protests during the coronavirus pandemic of 2020 was unprecedented and has focused attention on ensuring that the safety message on spectrum use is successfully com-municated to the public. The chapter by Dr Mary Longhurst and myself focuses on the campaign messages of both the anti-5G groups and industry. Why has the former been so successful and the latter failed to reach significant sections of the public?

This is a multifaceted question. At its simplest anti-5G protestors harness the power of social media and use many of the elements of successful campaigning – emotional messaging, powerful storytelling, the use of familiar narrative tropes. Most of the industry campaigns on 5G safety are low key and reactive compared to their marketing campaigns and lack the emotional punch and storytelling flair of the opposite side.

Generally, the science-backed industry argument is presented with insufficient thought for persuading the intended audience. But this need not be the case, as Nicky Preston's chapter shows. Voda-fone New Zealand mounted a humorous social media campaign to reassure the public about 5G safety which was influential in stem-ming the tide of arson attacks. The chapter shows the amount of work that went into identifying the target audience, establishing how they were receiving information on 5G, researching their level of knowledge on the subject, considering what would be an effec-tive way of persuading them and finding appropriate figureheads to present the message. It's the sort of work that would go into design-ing a marketing campaign for a new product and seems to have paid off handsomely.

Which points us to the complexity of communicating messages on 5G and health. It is influenced by the way government and public health bodies present their messages, and it has been affected by the changing media landscape in which social media has come to play such an important part. The mobile industry does not have direct con-trol over these, but it can control its wider communications strategy. As well as being consumers, the public are socially aware citizens, with increasing interest in health and the environment. A holistic communications strategy, covering all consumer touch points and recognising the importance of health messages, will make the public less likely to believe 5G conspiracy theories.

How Useful Is the Tragedy of the Commons Metaphor?

From the power of campaigns and messaging, we move to the influence of a single metaphor in the development of spectrum policy. The next chapter argues that the credence given to the "tragedy of the commons" is out of proportion to its credibility.

The idea that shared resources are inevitably inefficient because they are ultimately destroyed by over-use is much discussed in spectrum circles but is based on little evidence in the disciplines where it originated. In recent years, scholars studying the history and economics of land management have found few examples of the "puny and stunted cattle" which are supposed to result from the over-grazing of commons in the days before enclosure. They found many examples of complex networks of rules governing the public and the private which tried to balance a range of stakeholder needs to mutual advantage.

Sounds like a WRC? It certainly doesn't sound like the tragedy of the commons, but it does evoke the law named after the woman who won a Nobel prize for studying successful management regimes for common resources. Ostrom's law says that what is possible in practice must also be possible in theory: there is a challenge for students of spectrum policy.

Notes

1. E. Noam, 1998. Spectrum Auctions: Yesterday's Heresy, Today's Orthodoxy, Tomorrow's Anachronism. *Journal of Law and Economics*, 41(S2), pp. 765–790.
2. www.cnbc.com/2020/07/31/apple-surpasses-saudi-aramco-to-become-worlds-most-valuable-company.html.
3. EEB, 2019. *Cool Products don't Cost the Earth – Full Report.* www.eeb.org/coolproducts-report.

1

THE PAST, PRESENT AND FUTURE OF SPECTRUM AUCTIONS

MARTIN CAVE[1]

Contents

Introduction

The natural resource of radio spectrum was first discovered and used over a century ago. When its military significance was appreciated in the First World War, most countries chose to allocate and assign it centrally using administrative 'command and control' methods. The first comprehensive spectrum auctions, replacing administrative with market methods, took place 30 years ago in New Zealand, which was famous at the time as a pioneer in such approaches. The auction design adopted there was the 'second price auction': the band goes to the highest bidder but at the price offered by the second highest bidder. It led to unexpectedly low prices in some bands for which there was little competition but was otherwise considered a success.

In the USA, in the 1980s, the first alternative to administrative methods chosen was the use of lotteries, employed to assign the first batch of radio frequencies released for mobile services. The lottery offered 643 licenses[3] and attracted over 400,000 applicants, most of them small-time rent seekers intent on 'flipping' the licence, rather than ever building a network and providing a service. It was a calamitous process that lasted six years, delaying the launch of mobile services and doing nothing to serve the public interest.[2]

In the face of these contrasting results, in 1993, the US Congress allowed auctions to be held for non-broadcast spectrum. The 1994, simultaneous, multiple round auctions for 30 MHz lasted four months and netted USD 7.7 billion, which covered the historical costs of the Federal Communications Commission (the auctioneer) since its inception in the 1920s. Its success led US lawmakers to make auctions mandatory in 1997.

Thus, comparative hearings, also known as beauty contests, ruled the roost across much of the world until the turn of the century at which they gave way to competitive auctions fairly universally, at least in the burgeoning field of mobile communications, and because spectrum using wireless services were free from the 'natural monopoly' property which applied at the time to many wireline networks, a competitive auction usually embraced several licences covering the same service area. Thus, multi-bidder competitive spectrum auctions worked with liberalisation and competition in the services provided using the spectrum, to create a powerful innovative force.

Many early auctions, notably some of the 3G auctions held in Europe in 2000/2001, generated unexpectedly high revenues.[3] This did not escape the notice of finance ministries, which welcomed such inflows into their coffers. In this way, auctions soon became the norm, with only a few countries bucking the trend and maintaining 'command and control'.

This chapter takes the reader on an informal tour of the issues that arise with spectrum auction, since their inception 30 or so years ago, now, and in the future as the mobile sector, in particular, develops further. The tour covers the basic economic ideas behind auctioning natural resources; the policy objectives that auctions can pursue; how auctions can go wrong; the performance of 5G auctions; and some speculations about how they may develop in the future.

What Spectrum Auctions Can Do

We begin with an economics digression. The first prominent economist to support auctioning spectrum was R. H. Coase, British-born but working for most of his life at The University of Chicago. He is famous for the so-called Coase theorem, according to which if property rights are unambiguously assigned, then trade between the parties – market exchange – will allow resources to gravitate to their most efficient uses: "the delimitation of [property] rights is an essential prelude to market transactions."

In application to spectrum, in particular, he reasoned that, once there is legal certainty around a spectrum property right, it is possible to assign it, "employ[ing] the price mechanism, as this allocates resources to users without the need for government regulation."[4] This would mean that if A owns a spectrum licence which is now used to provide a broadcasting service, but B could use it more profitably to provide a mobile communications service, then a trade between the two can take place, which will benefit them both, and might also benefit the universe of downstream service users, whose comparative willingness to pay for the two services ultimately underwrites the comparative value of the spectrum to A and B.[5]

How better to get the process started with newly released spectrum than to have an auction of spectrum licences in which A and B can take on other bidders seeking the 'property right' and also to allow licence holders to trade their rights thereafter, in response to changes in tastes and technology?

Coase also believed that, with the market mechanism, the goal should be to maximise use of the spectrum, by owners who would then have the necessary incentives to deploy high quality and extensively available services.

This insight came to Coase in 1959. For this and a lifetime of other work, Coase was awarded the Nobel Prize in Economics in 1991.[6]

But it took policymakers in the USA 34 years to act on the message and implement spectrum auctions.

It is worth noting that it is fairly typical for a natural resource which is in plentiful supply to be allocated administratively by a government, probably at a zero or near zero price. But when or where it is scarce, that allocation becomes inefficient, and market methods are called upon.

Thus, we find that raw water in Southern California or Australia, where it is a scarce natural resource, is auctioned or traded, while water rights in the UK, where for now there is often too much water, are administratively assigned.

We have already noted, in the earlier discussion of Coase's ideas, the capacity for a competitive spectrum auction to direct spectrum to firms which can use it more efficiently and can, therefore, afford to bid more for it. This is the first and possibly the most important function a spectrum auction can discharge.

The second has more to do with who gets to benefit from the scarcity value of prime radio spectrum – say the 700 MHz band. If the government simply assigns it at a zero price to one or more operators, their shareholders will get access to a highly valuable and scarce asset for nothing and will convert it into high returns for shareholders and managers: they will get the so-called scarcity rents on the public asset.

But under an auction regime, that same scarcity value or rent will accrue to the government. Efficient operators should make normal returns, but the scarcity value would go to the finance ministry for use in various forms of public spending, transfers, tax cuts, the reduction of debt, relief of poverty, health and educational expenditure, etc. – as well as, unfortunately on some occasions, various wasteful or corrupt purposes. Bearing in mind that the cumulative global revenues from spectrum auctions to date is close to USD 1 trillion, it is of distributional significance where the money goes.

However, the versatility of auctions is not exhausted by greater efficiency and benefits for public spending. There are two further things more useful things which they can and have done.[7]

The auction rules can include caps on what the largest operators can buy in any auction, intended to restrain them from buying up all the available spectrum and thus eliminating or weakening their smaller rivals.

Finally, an auction can also achieve wider economic and social objectives by inserting them as conditions in the spectrum licence. For example, the regulator can require one or more operators to extend coverage to non-commercial areas. Or quantitative investment targets can be incorporated as licence conditions – although care must be taken to ensure that such obligatory inputs are chosen efficiently: it is usually better to specify mandatory outputs, not mandatory inputs. Or they could require spending on handset subsidies, or digital education – whatever can successfully be enforced as a licence term.

Any such condition, when it bites, will reduce auction revenues. It can be seen as a way of ploughing some of the auction revenues back into the sector – rather than diverting them to completely unrelated public purposes. Moreover, as the supply of these additional services is procured as part of a competitive auction process, it too can be directed towards the operators who can perform the tasks most efficiently. But this process breaks down when the auxiliary obligations are so granular that there is only one operator who can perform them: see later.

A recent Austrian example shows how these can be flexed in the course of an auction of 700 MHz and other bands, which concluded in September 2020.[8] Each of the three operators faced challenging initial coverage requirements and the outcome of the earlier stages of the auction was a combined bid of €300 million. In the final stage, the regulator took advantage of a capacity to flex coverage requirements further, by asking each operator to specify which sets of municipalities it would be prepared to cover in addition, in return for which reductions in its fee. (A process of this kind is known as a 'menu auction'.) The auction software then selected those bids which maximised coverage within a given budget. In this way, the regulator reduced its auction revenues by one-third, to €200 million. Note that this way of proceeding maintained competitive tension in the auction throughout.

How Can Spectrum Auctions Possibly Go Wrong?

Auctions can go wrong in a number of ways, some based on how firms behave (the demand side of the spectrum auction equation) and others based on how the regulator or government behaves (the supply side of the equation).

Auction Design Failures

Running a billion-dollar spectrum auction is a very complex business, requiring the drafting of meticulous rules to govern the behaviour of bidders and the auctioneer, the data to be exchanged by pre-specified deadlines and decisions about how to package the available spectrum. Another key set of decisions concerns whether to have a one-shot auction (requiring each bidder to send in a sealed bid, the winner being the highest bidder) or a multi-round ascending (or even descending) auction in which prices change in successive bidding rounds. If several lots are on sale, there can be a sequence of single unit auctions, or the bidding for all may take place simultaneously.

> Inevitably, not all auction designs have succeeded. A great deal of economic and mathematical ingenuity has gone into improving auction designs. And as already noted, in 2020, two US-based auction designers, Robert Wilson and Paul Milgrom, won the Nobel prize in Economics for their brilliant and inventive contributions.[9] But not all designers have been so smart.
>
> The most prominent examples of auction design involve the simultaneous auctioning of multiple lots – differentiated by band or geographical area. Increasingly operators need spectrum in contiguous areas to provide service to their customers, and also need spectrum in different bands to provide the right combination of coverage and capacity. They do not want the completion of an auction to leave them with a missing piece in their spectrum jigsaw puzzle. Two well known auction designer responses to these requirement are the Simultaneous Multi-Round Auctions (SRMA) and Combinatorial Clock Auctions.
>
> (CCA)

A recent paper investigated the relationship between the choice of auction design and realised prices using a data set covering 13,000 spectrum lots. It showed that the more sophisticated designs, like Simultaneous Multi-Round Auctions (SRMA) and Combinatorial Clock Auctions (CCA), which freed bidders from stranding risk, generate higher revenues, which are not huge but statistically highly significant.[10]

At this stage, there should be little excuse for choosing a failing auction design. But there is continuing scope for innovation. We close this brief discussion of auction design with an account of a particularly interesting auction designed by Paul Milgrom and others for the US Federal Communications Commission (FCC) and implemented in 2016/2017.[11] In most spectrum auctions, a given amount of spectrum is made available by a regulator: it is a 'one-sided' auction, without supply-side flexibility. Also, either under auction rules or in practice, all bidders plan to use it for a single purpose, usually mobile communications.

A more complex situation arose in the USA in connection with the 600 MHz band. It was used by broadcasters but coveted by mobile operators. The task was to test the extent of the willingness of some (but not all) of the broadcasters to sell their spectrum to mobile operators. In other words, a two-sided auction design was required.

We can characterise what is required for this process to work in the following highly stylised way. Put broadcasters and mobile operators with an interest in 600 MHz spectrum on either side of a room. An auctioneer suggests a trial price for a MHz of spectrum in the band; she adds up the quantities which broadcasters want to sell at that price and adds up the amount which mobile operators want to buy. Suppose broadcasters want less than mobile operators want to buy. The auctioneer raises the price and repeats the process. This goes on until a price is called out at which the market clears: the two quantities are the same. Sales then occur at this price.[12]

In fact, the US auction, known as the 2016/2017 'incentive auction', was hugely more complicated than this and was, of course, conducted electronically. But the success of this auction opens the way for the use of spectrum auctions in a wider range of applications than in the past.

Do Spectrum Auctions Raise the Price or Lower the Quality of Mobile Communications, or Discourage Investment?

We now conduct a thought experiment in which we notionally compare two states of the world. On the one hand, competition among bidders yields a competitive outcome. On the other hand, spectrum prices are arbitrarily lowered by a given amount. What happens to the extra profits thus generated in the sector? The context is assumed to be the commonly encountered one in which spectrum fees for the whole licence period are paid at the start. In other words, they are a sunk cost.

In order to investigate this issue, we have to cope with the issue of the direction of causation (also known as 'endogeneity'). Are service prices high because auction competition raises them? Or are auction prices high because customers value the services spectrum produces so highly. A study of data on auction prices, service prices and investment levels in 24 countries over the period 2005–2014 concluded (p. 363) is that

> after controlling for the potential endogeneity, spectrum fees do not seem to be correlated with spectrum revenues . . . the lack of significance of spectrum licences on revenue can thus be interpreted as a sort of 'correct' forecast of future revenues by mobile operators in the licence paid: the cost of the licence already incorporates the future increase in revenues but it does not have any further incremental effect on them.[13]

More recent work for the GSMA, the body that represents the mobile industry has, however, found associations between high auction prices and slower roll-out and poorer quality of services.[14] In keeping with the argument made earlier, over the ability of the regulator or government to plough some of the auction proceeds back into the sector by specifying and enforcing spectrum licence conditions relating to coverage, speed, quality of service, or even investment levels upon all operators, or an individual operator, a solution to such issues might well be found within the auction design process itself.

This is easier said than done, of course, since such licence conditions must be established at the start of what might be a fairly lengthy licence period, although they can be time-variant.

Overbidding, or the Winner's Curse

Another possible problem with auctions is that they may generate overbidding, consequent on a systematic tendency for at least some operators to harbour optimistic projections of the future and thus bid more than the spectrum is worth. When this occurs, the over-bidder might seek to retrieve the situation by raising service prices. But this would require similar action by all suppliers in the market, which is not likely to be forthcoming. Also, bidders for mobile spectrum are typically large international operators with access to appropriate advice and familiarity with the risk of the so-called winner's curse, although the curse may have prevailed in the German 5G spectrum auction described in the following, where the immediate reaction of winners after the event was somewhat disconsolate or even bitter.

Supply-Side Restrictions on the Availability of, or Competition for Spectrum

The last problem we discuss is particularly troubling and intractable. It arises when Governments restrict spectrum availability to bid up spectrum prices: in other words, when they exploit their monopoly power to drive prices up by restricting supply. As part of this, they may delay the release of spectrum to create pent-up demand for services and push up prices further.

An equally crude method is to sell all the available spectrum in one lot: the highest bidder is thus gifted market power in the downstream market, which – when valued by the method described earlier – generates a larger prize for that bidder.[15]

Each of these outcomes has adverse effects on coverage, quality and price. They lower take up of the service, restrict connectivity and stunt economic growth and the higher government tax receipts which go with it. If this happens, there is an association between higher spectrum prices and higher service prices, but the former are not causing the latter. The causation is the other way round: it is the expectation of the high profits caused by manipulation of the spectrum market, which generates the higher bids.

By doing this, a government shoots the population of its country in the foot by putting up connectivity prices, restricting economic growth and reducing tax revenues. However, finding a universal solution for this problem is extremely difficult.

What Do Recent 5G Auctions Suggest about the Current State of the Market?

The radio spectrum is non-exhaustible: it does not wear out. The sooner it starts being used, the better. Ronald Coase emphasised in

1959 that the more if it there is available, the cheaper it is likely to be, and the more it will be used to end-customers' benefit. But as the first wave of 5G auctions played out in Europe and the USA, the amounts of spectrum supplied to those two marketplaces were highly contrasted.

The US President has declared that "the race to 5G is a race that America must win" and to give its industry the best shot, the FCC began flooding the market with spectrum, as spectrum licences are being sold at historically low prices. The FCC Chairman Ajit Pai remarked that its "we're taking an aggressive, all-of-the-above approach: we're freeing up high, mid, and low-band spectrum for 5G. Looking high we've . . . made available to the private sector a combined 1,550 megahertz of spectrum." And they were not stopping there. A third spectrum auction in 2019 would be the largest in American history, releasing 3,400 megahertz of spectrum into the commercial marketplace.[16] That process continued throughout 2020, with a large auction planned in 2021 for 280 MHz in the 3.7–3.98 band.

Compared to the American flood, the pace of European spectrum supply was more of a trickle. The primary 5G pioneer band in Europe is 3.4–3.8 GHz, and a deadline of end-2020 was set for member states to "reorganise and allow the use of sufficiently large blocks of the 3.4–3.8 GHz band."[17]

In addition, member states were supposed to assign 700 MHz, 3.5 GHz and 26 GHz licences before the end of 2020, but the vast majority of them have failed to do so.[18] Greece now joins Italy and Finland as the only nations which have successfully awarded licences in all three bands by 31 December 2020, in line with the rules laid down by the European Commission.

The European Commission expressed its impatience with progress in its Recommendation on deploying very high capacity network of September 2020.[19] This included fairly precise instructions for authorising spectrum in the 700 MHz, 3.4–3.8 GHz and 24.25–27.5 bands.

It is not difficult to identify possible reasons for this dilatoriness. First, the scope for litigiousness over auction rules by rivalrous operators is substantial. Second, spectrum regulatory agencies are bureaucratic bodies which are not accustomed to exhibiting agility. And third, governments may want to delay matters to enhance revenues.

Europe's track record on meeting such deadlines is a cause for concern and the continent can ill afford a re-run of the 4G race that saw other regions accelerate out of the blocks and assume an unassailable lead. Whereas the USA had auctioned off 84 MHz of digital dividend spectrum in 2008, by 2013 only 9 of 27 EU member states had come good on their spectrum release commitments and consequently the USA had ten times as many 4G subscribers.

Table 1.1 shows the outcomes of 5G auctions in the 3.4–3.8 GHz band in the EU 2017–2020, in terms of the standard metric of Eurocents/pop/MHz.

We draw attention to three of these results in particular:

- In Italy, the auction generated very high prices. This has been attributed to packaging of the 200 MHz of spectrum into two blocks of 80 MHz and two of 20 MHz. With four operators bidding, including a new entrant bidding for the first time, this sets up intensive competition for larger blocks, which may have been seen as a route to gaining some market power.[20]
- In Germany, the auction took 52 days with 497 rounds of bidding. The final round had the same allocation of spectrum

Table 1.1 Outcome of Selected 5G EU Spectrum Auctions in the 3.4–3.8 GHz Band, 2017–2020

MEMBER STATE	AUCTION DATE	PRICE (€CENT/POP/MHZ)
Czech Republic	July 2017	1.8
Ireland	May 2017	4.7
UK	April 2018	13.0
Spain	July 2018	4.7
Finland	October 2018	3.6
Italy	October 2018	35.9
Austria	March 2019	5.8
Germany	June 2019	16.8
France	October 2020	13.4
Greece	December 2020	4.7

blocks as round 191 but with highly inflated prices. A new entrant had come into the market, spending over €1 billion, and some of the band had been withheld for assignment not to mobile operators but to verticals, which self-supply their own spectrum rather than use bands licensed to mobile operators. The coverage and speed obligations were intense. The high prices attracted unfavourable comment from operators, one saying: "The price could have been much lower . . . network operators now lack the money to expand their networks." What drove the prices up was the behaviour of mainly highly experienced operators engaged in a familiar competitive process. It may be a recognition of the previously mentioned winners' curse.

- In France, the regulator adopted a two-stage process. In the first stage, the regulator assigned 50 MHz to each of four operators at a price of €350 million; roll-out of 5G services across 10,500 sites was associated with this tranche. In the second stage, the four operators bid competitively for 110 MHz. It was thus an imaginative combination of a 'command and control' approach, with a market-structure preserving offer to each firm of a uniform amount of spectrum for a set price, combined with a more standard 'top-up' auction, which was in practice dominated, as expected, by Orange, the largest provider. The set price per MHz was €7 million in the first stage, and the market-clearing price was more than €15 million in the second phase. At first sight, this might suggest an element of below market-rate pricing in the first phase, but it must be remembered that the first stage carried substantial roll-out obligations.

However, we cannot fail to mention the countries that have not held spectrum auctions but have persisted with the use of administrative methods. The two most prominent are Japan and China.

Both China and Japan do not charge upfront fees for spectrum but instead judge whether the deployment commitments of firms match the country's policy objectives and assign spectrum accordingly.

Both countries currently enjoy more expansive connectivity than their global peers. Measured by the number of installed 4G base stations per person, mobile networks in China and Japan are around four times as dense as those in larger European countries and the USA.

Although network densification may be attributed to multiple factors, not least favourable planning rules and low passive infrastructure access costs, an operator's balance sheet that is not stretched through excessive spectrum payments will play a role. As noted later, network densification is a key element for the successful realisation of an expansive version of 5G and should be an increasingly important consideration for all spectrum assignment policy objectives and be reflected in how spectrum auctions are designed.

The mobile sector contexts in China and Japan are quite different. In relation to spectrum auctions, this was brought home to me when I attended a meeting in May 2018 in Beijing at the Advanced Forum of Spectrum Resource Marketization hosted by the Radio Association of China. Following a discussion of spectrum auctions in various countries, I was asked whether it would be practicable or desirable to hold a spectrum auction in China involving the three mobile companies in public ownership.[21]

Industry–government relations in Japan tend to be closer than in Europe and the USA. And historically a degree of carefully negotiated consensus concerning spectrum assignments and appropriate operator responses has emerged. This has included arrangements for spectrum to be made available to a new entrant.

The Future of Auctions

We end with some brief thoughts about the future of spectrum auctions. I have argued elsewhere that it is striking how stable the market structure of the mobile sector has been since the 1980s, while four generations of technology were assimilated and voice-only services gave way to data and the smartphone.[22] Jurisdictions tend to have broadly similar numbers (often three or four) of distinct, vertically integrated operators.

It has been noted that 5G may challenge this stability: not so much in its more limited manifestation as a faster and cheaper version of 4G, specialising in enhanced mobile broadband, but in its more expansive form in which very fast and low latency communications capacity is going to be available everywhere, and employed in 'verticals' not so far fully penetrated by digitalisation and connectivity, such as connected cars, advanced manufacturing, e-health – in short, almost every aspect of consumption, production and social activity.[23]

This will involve the densification of networks, which may bring into question the survivability of even the increasingly shared networks observable today. Further regulation to control the market power of the smaller number of operators may be needed, and the scope for running spectrum auctions with very few bidders may diminish. This may endanger spectrum auctions.

However, individual 'verticals' may prefer to self-supply their own connectivity using their own spectrum, rather than outsource this to mobile operators. This might enhance the competitive nature of auctions.

Alternatively, such a vertical, in either the public or private sectors, may choose, in preference to requiring their customers to buy an individual retail communications service from a mobile operator, to buy a bespoke slice of a network and bundle it with its own retail provision to its customers.

These developments are already foreshadowed in recent 5G auctions. Thus, the French 5G auction described the aforementioned provisions that require spectrum licensees to guarantee connectivity to industry verticals, to offer network slicing tools by 2023 and to host any mobile virtual network operator (MVNO).[24]

In the German 3.4–3.8 GHz auction, the regulator (BNETZA) reserved one-quarter of available spectrum for verticals. Since 2019, 91 closed user group assignments have been made available to local industry. Each user must negotiate local arrangement with its neighbours. A fee of €31,000 is charged per square km for access to 100 MHz over ten years. Applications for spectrum on a similar basis in the 24.25–27.5 GHz band will be accepted from 2021.[25]

These, combined with localised sharing of nationally assigned spectrum with verticals, are potentially important developments.

Conclusions

Over the past 25 years, spectrum auctions have become a standard means of assigning high-value spectrum. They have offered a competitive means of allocating spectrum efficiently among operators, at a time of large increase in spectrum scarcity. It does not seem likely that the allocation system they replaced – 'beauty contests' – could have achieved this goal.

Auctions allow the government, not the operators, to capture the rents associated with that increased spectrum scarcity and use them for various public policy objectives, critically including policy objectives pertaining to the mobile sector itself. Thus, spectrum auctions have also had grafted on them, in the form of spectrum caps and coverage obligations, by means of which greater downstream competition and wider deployment of networks can be gained at the cost of some of the government revenues achieved. Obligations are an increasingly important trade off to consider in wake of the potential positive external benefits that could be associated with the expansive version of 5G.

This would be a more constructive approach than a poorly constructed 'command and control' approach, in which different operators seek to deploy their bargaining skills, lobbying skills and private knowledge to influence the outcome of an administrative process. The auction approach retains its traditional competitive advantages, which have brought considerable benefits to the countries that have employed them. At the same time, auctions can be adapted to the new circumstances, and the balance between government revenue and increasing deployment can be flexed. This is likely to be a better course of action than 'throwing the (auction) baby out with the bathwater.'

To promote the reach, availability and quality of connectivity, some auction best practice approaches have emerged:

- Align spectrum pricing with policy goals, for example, by including coverage and other obligations in the pricing objective.
- Package spectrum into small lots to support rational and competitive bidding
- Avoid sealed bids, reduce complexity and ensure transparency.

- Avoid the temptation to restrict supply of spectrum in pursuit of short-term fiscal goals.
- Think carefully before removing spectrum away from competitive assignment by setting it aside for particular users or uses. Such approaches have generally not been successful. But when a major structural break in spectrum arrangements is on the cards (as may be the case now in relation to spectrum acquisition by verticals), some positive steps of encouragement may be justified after a clear-eyed analysis both of risks and of alternative options.
- Maximise licence duration to increase the asset life and investment horizon and set out renewal criteria to reduce uncertainty around investment.
- Ensure predictability and allow for budget planning by providing a clear roadmap of spectrum assignment over the medium to long term.
- Allow secondary trading of licences along with the ability to sub-lease amounts of spectrum to third parties.

Notes

1. This paper draws on some joint work with Gabriel Solomon, but the author retains responsibility for its contents.
2. For a very insightful and enjoyable history and analysis of US spectrum policy, see T.W. Hazlett, 2017. *The Political Spectrum*. New Haven, CT, USA: Yale University Press.
3. Thus the $69 billion revenues (in 2021 prices) raised from the 2000 3G spectrum auction in Germany was unsurpassesd until 2021, when the auction of the 3.7–3.98 GHz band in the US yielded $80 billion. T. Youell, January 18, 2021. US Sells C-Band for $80.9 Billion. *Policy Tracker*.
4. R.H. Coase, October 1959. The Federal Communications Commission. *Journal of Law and Economics*, 2, pp. 1–39.
5. It is a key practical point, bearing in mind interference between radio signals, that that neither A nor B may trespass by way of interference on the defined property rights of adjoining band user C.
6. I note below that 29 years later in 2020, two other US-based economists jointly won the same Prize for their work on the design of auctions, including particularly spectrum auctions.
7. These are further outlined in M. Cave and R. Nicholls, 2017. The Use of Spectrum Auctions to Attain Multiple Objectives: Policy Implications. *Telecommunications Policy*, 41(5), pp. 367–378.

8. M. Marti, 16 September 2020. Austria Completes Delayed 5G Tender. *Policy Tracker*.

9. See the Royal Swedish Academy of Sciences, Nobel Prize in Economics, 2020. *The Quest for the Perfect Auction*, pp. 4–6. www.nobelprize.org/uploads/2020/09/popular-economicsciencesprize2020.pdf.

10. P. Koutroumpis and M. Cave, June 2018. Auction Design and Auction Revenues. *Journal of Regulatory Economics*, 53(3), pp. 275–297 See: link. springer.com/article/10.1007/s11149-018-9358-x

11. FCC, 2017. *Broadcast Incentive Auction and Post-Auction Transition*. www.fcc.gov/about-fcc/fcc-initiatives/incentive-auctions.

12. Because broadcasting and mobile eservices interfere with one another, the broadcasters then had to be reassigned frequencies which packed them into the lower part of the shared band.

13. C. Cambini and N. Garelli, 2017. Spectrum Fees and Market Performance. *Telecommunications Policy* (41), pp. 355–366, at p. 363.

14. GSMA, September 2019. *The Impact of Spectrum Prices on Consumers*.

15. Some observers have noted a similar but more muted effect of this kind in the 2018 Italian 5G auction described below.

16. docs.fcc.gov/public/attachments/DOC-359335A1.pdf.

17. See: Article 54 of the European Electronic Communications Code.

18. M. Marti, 30 December 2020. 24 Out of 27 EU Countries are about to Miss Spectrum Deadline. *Policy Tracker*.

19. COMMISSION RECOMMENDATION of 18.9.2020 on a common Union toolbox for reducing the cost of deploying very high capacity networks and ensuring timely and investment-friendly access to 5G radio spectrum, to foster connectivity in support of economic recovery from the COVID-19 crisis in the Union.

20. M. Marti, 3 October 2018. Italy's 5G Auction is Over. *Policy Tracker*.

21. If appropriately chosen competitive objectives and incentives were imposed on the companies in public ownership, it might be possible in principle to replicate an auction outcome similar to that observed with investor-owned operators, but for various embedded cultural reasons in practice it would be difficult to achieve this outcome.

22. M. Cave, 2018. How Disruptive is 5G? *Telecommunications Policy*, 42(8), pp. 653–658.

23. See W. Lemstra, 2018. Towards the Successful Deployment of 5G in Europe: What Are the Necessary Policy and Regulatory Conditions? *Telecommunications Policy*, 42, pp. 587–611.

24. M. Marti, 2 October 2020. France Auctions the 3.5 GHz Band for €2.8 Billion. *Policy Tracker*.

25. T. Heutmann, 8–10 December 2020. *Frequencies for Industry in Germany*. ITU & CITC Workshop: Radio Spectrum for IMT-2020 and Beyond.

2

Spectrum Auctions

From the Advent of Auctions to Post-Pandemic Policy Imperatives

GÉRARD POGOREL

Contents

DOI: 10.1201/9781003156765-3

2.1 Introduction

Economic approaches and valuations have played an increasing role in spectrum policy and management since the last decade of the 20th century. Increased competition in telecommunications services in the past two decades has generated considerable benefits for citizens, consumers, industry and governments in Europe and all over the world. In 2020, the network operation performance of telecom operators during the pandemic crisis has been justly commended. However, the shortcomings of connectivity and data speed in some areas, the difficulties it has provoked for distance learning, remote working, question the rules and rationales which have prevailed in telecommunications services regulation and policies. Overcoming the excessive reliance on abstract economic reasoning and enhancing the status granted to essential social and public policy objectives in policies and procedures are evidently the order of the day.

This chapter analyses the role spectrum assignment policies have played in this evolving picture and explores in what way the design of spectrum awards from now on should give centre place to promoting ubiquitous infrastructure deployment and maximising "social value".

2.2 From Techno-Centred to Market-Oriented Management: The Turn of the Century Change in the Radio Spectrum Paradigm

Until the 1990s, technology and engineering considerations have prevailed in the management of the radio spectrum. The last decades of the 20th century saw growing concerns that the promise of breakthroughs in communications technologies were being hindered by the powers of legal monopolies. Ronald Coase seminal 1959 paper[1] had paved the way, advocating that spectrum assignment should be based on market and pricing mechanisms, where property rights are assigned with the objective of maximising output from a scarce resource. From the 1990s, the perception of a speeding up of the pace in technology innovation, growing attention to competition economic considerations,[2] and the realisation of potential windfall gains for Government budgets have combined to create a shift to auction-based frequency assignment to mobile services to overcome the lack of transparency of administrative procedures in selecting the competing licensees. Auctions were adopted around the world following the USA's successful example in 1994. They were considered more transparent, quicker, and removed the need for governments to pick winners.

This logic helped open the telecommunications industry to competition and brought about a change in business culture and practices. The belief that "market mechanisms" in spectrum management were an optimal means of solving assignment problems and promoting the industry and the economy was widely shared and delivered the communications revolution we all benefit from today. The main milestones of this initial shift to market mechanisms and improved spectrum assignment procedures were the 1998 European Commission Green paper,[3] the 2002 Federal Communications Commission (FCC) spectrum policy task force report[4] and the 2002 UK Martin Cave independent Review of radio spectrum management.[5] On a more mundane, but no less significant level, it is said the French Minister of Finance at the time, when learning about the proceeds for the public budget of the UK and German auctions of something very arcane called radio spectrum, had "stars in his eyes".

At its peak, the market approach to radio spectrum allocation ambitioned to apply to most, possibly all, areas of frequency bands. In the most extreme scenario, once put in place exclusive property rights, auctions and trading would take charge of the whole of frequency allocations, making planning redundant. Considerations about monetising TV broadcasting bands succeeded in the USA, with incentive auctions on broadcasting frequencies. In the UK, an attempt at monetising civil aviation frequencies floundered. The defence spectrum in the UK is a case in point, as the British government committed to allow additional military expenses in proportion to released spectrum, both movements taking place, however, within public pockets. Eventually, the shift to auctions was mostly limited to mobile. It does not mean that economic considerations are not absent in other commercial areas of frequency allocations, like satellite or PMSE, but they remain implicit and in the background.

Over the last two decades, spectrum policy has been widely governed by the same high-level, generic, economic principles, we can summarise as follows:

A   Auctions are the optimal mechanism not only to determine a spectrum price but also to align private and public objectives, as by actively bidding, operators demonstrate their willingness to invest in network deployment.

B   By licensing a public resource, governments should aim to maximise the price or, more correctly, to use market mechanisms to determine the amount paid by buyers that balances supply and demand of radio frequencies.

Principle A is based on the fundamental Econ 101[6] assumption that market mechanisms consistently provide the optimal outcome in resource allocations. However, evidence suggests that potential shortcomings with this assumption abound in relation to the deployment of communications networks. Governments and National Regulatory Authorities (NRAs) have waged an ongoing fight against natural monopoly resurrection tendencies. Moreover, competition alone has not been sufficient to achieve specific public policy objectives. Almost all countries have taken measures to complement market mechanisms and provide public policy defined levels of services not only in rural, low-density areas but also in many urban districts, with even some single streets in big cities having been designated as "market failure areas".[7]

Principle B applies Econ 101 to the use of public resources. It is a bold assumption that for the state to handle public funds optimally, it must behave as a private agent in the management of its proprietary assets. This argument has a strong element of opportunism in times when tight government budgets are badly in need of income. The evidence, however, shows that discrepancies arise between misguided public management procedures of spectrum assignment and the realisation of public objectives in industry and market growth.

The high-level, often repeated, argument that auctions are an effective process to select those operators that can use the frequencies most valuably is as hard to verify as it is to falsify. But once the assumption is accepted (Econ 101), auctions become synonymous with assignment.

Auctions are such an "exciting"[8] game, economic and social consequences are conveniently set aside. The higher the fee, the more successful the performance. Auctions bring emotions, if not reason, into the otherwise dreary world of radio spectrum.

Spectrum auction studies in the first decade of the 21st century have failed to address what they were meant to research, i.e. the impact of spectrum assignment design on industry development, competition and contribution to the overall economic growth. The means have superseded the aims, and a "successful" auction in the literature has been taken as one achieving high fees rather than positive social returns.

Ironically, the growing popularity of spectrum auctions took place at a time when once triumphant Econ 101[9] was being abandoned, or at least side-lined in most areas of applied economic analysis. With the current generation of economists, general market equilibrium considerations and the reliance on generic microeconomics have lost ground to careful analysis of specific concrete market situations.[10] Over the last two decades, we have witnessed a growing emphasis on applied economic analysis with an increased focus on economic realities and a move away from high-level non-falsifiable theories. This is now also happening in the debate on mobile industry economics. Spectrum economics has lagged in this respect but is now catching up and several recent studies on the outcomes of spectrum assignments provide a clearer, pragmatic view of the results of auctions.

2.3 Studies Show Spectrum Auctions Have Not Delivered the Best Possible Benefits from Mobile

Recent studies suggest why and how the conventional approach to licensing spectrum has fallen short of connectivity policy objectives. There is a growing body of evidence that spectrum auctions in their current configuration not only fail to stimulate network investments but also hinder them.

In their (2017) paper[11] *The effects of spectrum allocation mechanisms on market outcomes: Auctions vs beauty contests*, Kuroda Toshifumi, Baquero Forero and Maria Del Pilar compare the evolution of market outcomes in 47 countries after the assignment of mobile spectrum by auctions and beauty contests held from 2000 to 2008:

> Traditional auction theory predicts the merits of auction versus 'beauty contests'. . . . We employ two semi-parametric estimators to determine the treatment effects and find that 3G mobile phone penetration rates among auctioning countries are 1.04–8.95% lower. Findings suggest that auctions used to raise public revenues not only transfer profits to the government but also sacrifice consumer surplus.

A research report *Effective Spectrum Pricing* by GSMA and NERA (2017) concludes[12]: "Statistical evidence shows the impact on consumers and links high-price outcomes (in auctions) with:

- Lower quality and reduced take-up of mobile broadband services.
- Higher consumer prices for mobile broadband data.
- Consumers losing out on economic benefits with a purchasing power of an estimated US$250 billion across 15 countries where spectrum was priced above the global median – equivalent to US$118 per person.
- Lower spectrum input costs are linked to greater price competition and higher usage."[13]

A study by *PolicyTracker*, LS Telcom and VVA (2017) for the European Commission[14] finds that "the grouping with the highest auction prices also had the poorest network availability. This questions the common view that operators who pay high prices for spectrum must invest in their networks to make this money back". Countries where operators have paid the most for spectrum over the past decade showed the worst 4G network availability.[15]

Cambini and Garelli[16] (2017), in a research covering 24 countries in the period 2005–2014, found that spectrum fees and availability do not have a significant impact on operators' revenue and investments. "The analysis provides evidence that spectrum availability and fees are not significantly correlated with mobile industry revenues suggesting that market expectations to extract additional revenues from the mobile service following new spectrum auctions are likely not to be respected". This would indicate that spectrum fees are treated as sunk costs by operators and, therefore, have no impact on investment and pricing decisions, thus voiding the incentive role of auctions on investments.

Hazlett, Munoz and Avanzini (2012) had demonstrated that efficiencies associated with retail services in mobile markets are about 240 times as large as those associated with licence revenues.[17]

These empirical studies confirm the analytical assumptions by Pogorel and Bohlin (2017) that spectrum auctions aimed at high spectrum fees do not serve to stimulate investments and network deployment.[18]

In a classic example of the right hand ignoring what is being done by the left hand, some branches of government, or agencies in charge of licensing, have tended to focus exclusively or primarily on maximising the fees they can derive from the spectrum auction procedures. Only secondary attention was being paid to the now widely observed limitations of this policy tool in achieving broader policy objectives.

It is now clear at both conceptual economic analysis and empirical levels that local market mechanisms in the form of spectrum auctions do not align with industry development objectives. Public policy objectives, which are central to the issue of non-discriminatory access to spectrum, have only partially been achieved. Some reprioritising and realignment of budget criteria with the number one objective of public policy "social value", considering business profitability, are required.

A functioning generalised connectivity is now recognised as a top priority in the hierarchy of public objectives, also a pre-condition to reap the best possible amount of resources in the medium term for the government budget.

2.4 Revisiting Spectrum Assignment – from Spectrum Value to Social Value

The pandemic crisis has already had an impact on the way we look at spectrum and wireless services. The discourse on the "value of spectrum", which was already losing steam, is now depreciated given the priority to be given to the "social value" of urgent connectivity. More than ever, spectrum has no intrinsic value. Its value resides exclusively in the contribution its use makes possible for society and the economy.

It is now widely recognised the principles that governed spectrum assignment in the last century and the beginning of the 21st century are no longer valid and should be re-visited, in particular, in the light of the generalised connectivity imperative which emerged from the pandemic crisis. We see residual tensions between NRA positions on coverage and certain government departments' focus on revenue. There is also scope for divergence between government ministries, with, for example, finance and regional development departments promoting contrasting views on whether to prioritise revenue maximisation for debt relief or deployment for economic recovery.

It is also to be noted that some 5G spectrum auctions have gone haywire due to the simple fact that 3G/4G mobile operators can't afford not to get 5G spectrum and be kicked out of the market. This forces them into an uncontrollable bidding process, devoid of business perspectives.

The blind faith in a solid relationship between spectrum auctions and mobile deployment has given way to a wide array of hybrid criteria procedures that have in common the trade-off between frequency fee requirements and territory and population coverage. Social value spectrum assignment scenarios are now implemented to balance the requirement to efficiently use a limited public resource, with the equally important objective of maximising the economic and social benefits that flow from investments in mobile network infrastructure.

One difficulty, in the past, was that social and economic impacts, whatever the efforts devoted to analysing them, are conceptually and practically hard to precisely measure and quantify.[19] It is true if all the growth percentage points supposed to derive in the last three decades from computers, networks and digital progress, in general, were aggregated, our economies would fly at space shuttle speeds. However, from 2020 to 2021, the penny has dropped and the setting of urgent coverage and deployment network objectives in the assignment procedure is vindicated.

2.5 Current Spectrum Licensing Procedures: A Survey

Governments and regulatory authorities implement a wide array of spectrum licensing[20] procedures embedding positive short-term and long-term impacts at telco, industry, government budget and macro level in the terms of the licensing process. This allows operators to know precisely what is expected in the terms of their licence and to define their business model and strategy accordingly. Competition stays rightly at the heart of telecommunications industry dynamics with competitive bidding between firms central to effective assignment processes.

2.5.1 Improved Auctions: Frequency Fees with Coverage Obligations

The latest generation of assignment procedures has given up pure auctions on fees and included more precisely determined and stringent obligations, defined as population and/or geographic coverage commitments. Auctions with coverage obligations have been frequent in spectrum assignments in Europe: 25/25 assignments in the 800 MHz band and 12/23 in the 2.6 GHz band (Magi).[21] The 2021 C-Band auction in the USA also includes phased "construction requirements".[22]

NRAs have come up with provisions attaching coverage obligations specifically to some bands or lots only. Sweden's regulatory authority, the PTS,[23] in a 700 MHz assignment of 2011, attached coverage obligation only to one of the six lots, so that only one operator was obliged to meet the requirement. The DEA in Denmark has retained the same asymmetric requirement in its 2018 procedures for the 700 MHz and 900 MHz band auction.[24]

One approach that is sometimes considered, but rarely tested, is to put auction proceeds into a fund ("USO-type"). This fund could then be actively used, through public purchasing, to cater to prioritised societal and political needs, including redundancy in networks and coverage in rural and remote areas. All, of course, subject to state aid controls. The model implemented in Sweden in 800 MHz is in line with this approach.

Many NRAs, based on their appraisal of the public interest, have opened auctions with dual objectives: combining 90–95% coverage conditions and a spectrum fee auction. By doing so, they entrust the bidders with a somewhat conflicting commitment. The payment of the fee will make it more difficult to invest in the network. This is a risk shared by the NRAs and the operators, but, in the end, it can always be said that the fund devoted to the spectrum fee could have been put to better use allowing quicker deployment of the network. Alternatively, in cases like early stage 5G deployment, where technology and economic risks and uncertainties are high, certain NRAs might not want to pre-define coverage obligations. Possibly, the assignment mode would warrant from the bidders' substantial but more progressive investment steps. The required coverage might be different for different frequency bands, taking into account technical complementarities, for instance, between sub-1 GHz bands and those above 3 GHz.

An interesting twist in the coverage/investment combination was featured in a 2018 Danish award with a provision that "Winning bidders will have the option of bidding for extra coverage obligations in exchange for a reduction in their licence price".[25] The objectives proposed by the bidders would presumably be expressed in quantitative terms, mirroring the political aspirations of connectivity, coverage, quality and speed.

This scheme opens the possibility for the regulator to avoid the dilemma between pre-defined and operator-defined objectives. It provides an interesting way for operators to adjust their investment objectives within an auction procedure.

2.5.2 Assignments Prioritising Investment and Coverage Commitments

In assignments prioritising coverage obligations, operators are in the driver's seat as the bidders' commitments are left to business strategy considerations.

2.5.2.1 Negotiated Frequencies-for-Investments

The widely publicised ARCEP "New Deal" in 2018 has covered the renewal of 4G licences. It waived auction-level frequency fees as a counterpart to committing operators to an intensive and accelerated country coverage program, the ultimate impact of the plan being more investment over the next three years than in the last fifteen years.

This approach has so far only been implemented in France and partially in Sweden. The challenge is the competition angle. The deal is limited to existing licence holders. Can new entrants be admitted to the negotiating table? How can they be selected? To combine the competition imperatives with the "New Deal", a two-stage process could be implemented:

- Select one or more new entrants through tender
- Negotiate their investments

2.5.2.2 *Assignments with Alternative Levels of Coverage Commitments*

As opposed to auctions on frequency fees with coverage obligations, the commitments of bidders in assignments with pre-determined alternative levels of coverage are of their own choice. Unlike the negotiated frequencies-for-investment ("New Deal") model, the process is competitive from the start.

Such an option has been implemented in 2019 for the 5G spectrum assignment in Austria. The regulator RTR proposed different coverage levels obligations ranging from virtually complete population coverage by 2025 or 98% coverage of certain highways and railway lines, plus the requirement attached to 700 MHz licences to cover at least 900 out of 2,100 underserved rural municipalities by 2027. The Austrian 5G spectrum assignment doubled the minimum requested number of underserved communities covered using 700 MHz spectrum reaching about 81% of all the underserved municipalities in Austria.[26]

To conclude on assignment procedures, let us not forget that, after some debates among Government departments, Japan, which managed to introduce a former mobile virtual network operator (MVNO) as a network-based new entrant in mobile by administrative decision, choose to stick to an efficiently performed administrative assignment of 5G frequencies that enabled the country for a headstart in network deployment.

2.5.2.3 *Pricing the Use of Spectrum as Limited Public Domain*

In investment-centred assignment procedure, frequency fees do not play the central role. The procedure must nevertheless determine what charges should be paid for the use of spectrum as a limited resource. What frequency fee should be paid to the government? We could consider various methodologies:

- Percentage of investments/deployment commitments
- Percentage of expected income
- Pre-defined flat fee

The terms of payment could be:

- Upfront
- Annuity instalments
- Spectrum annual fees

Alternatively, and more radically, the spectrum fee could be waived or limited in order to favour investments. The government will benefit from increased incomes of citizens and all industries and from the corresponding tax receipts over time. This focus on investments also has positive impacts on R&D, technology and standards.

The benefits from extended and accelerated network deployment will accrue to the public budget as well, through successful economic activities. To permit this expansion, it will be increasingly important due for the public sector and local municipalities, to facilitate, and not hinder the necessary infrastructure deployment.[27]

Ultimately capital expenditure (CAPEX) and operational expenditure (OPEX) amount to an investment equation: frequency fees, network deployment and coverage obligations of different natures, per frequency, per geography and over time. The risks associated with the various dimensions of the business activities being developed add up to a simple familiar investment calculus. Two factors can disrupt this: the reserve price set by the NRA in the auction and the auction process itself.

2.5.2.4 Monitoring and Compliance

Another key issue is the compliance of bidders in the implementation of the deployment commitments in terms of coverage, speeds, consistency, schedule, in their bids. A major risk is the potential divergence between ex-ante commitments and ex-post outcomes. While pure auctions are based on fire and forget ex-ante expectations, all investment incentive auction design relies on carefully designed rules of behaviour and follow-up monitoring. To make sure that the investments the operators have promised are indeed taking place, institutional arrangements should be designed to ensure the compliance of connectivity outcomes to commitments and to cope with potential shortcomings.

One possibility is to put the investment proceeds in escrow to be released to match deployment by the operators. Sweden's PTS did this in 2011. Keeping investment budgets in a fund, with adequate yield, has the advantage of reversing the burden of the proof: it is up to the operators to demonstrate they have complied with their commitments, not to the NRA to prove they have not.

In the end, the task of monitoring the achievements of the selected licence holders commitments will not be much different from what is currently performed by NRAs. It is indeed delicate, but NRAs have been dealing with it in many instances. Some flexibility should indeed be allowed on investment plans given changing economic conditions. A degree of flexibility of investments in specific bands is warranted: the commitments cannot be band specific over the extended period.

The issue of returning unused or under-used frequencies if the commitments are unfulfilled must be considered, as is the case with past spectrum assignment procedures.

2.5.3 Direct Assignment of Spectrum for Vertical Industries

Much is expected to form the contribution of 5G to Industry 4.0. The regulatory framework for vertical industries access to radio frequencies has been the subject of innovative measures in most of the major industrial countries in Europe, in Japan and in several Asian countries. Manufacturers, or local players, can benefit directly from local licences in certain pre-defined frequency bands, at relatively low fees. This has been the case, for instance, for around 100 sites in Germany at the end of 2020. To provide connectivity to industrial production, various network solutions are made possible, ranging from the use of operators' public networks, the deployment of private networks (self-deployed or via third parties), to hybrid solutions, combining elements of private and public networks.

Many players are thus involved in the ecosystem of 5G industry, expected to making it dynamic. Alongside the manufacturers themselves, we find companies specialising in industrial automation, digital service companies, suppliers of communication equipment, mobile communications operators, which manufacturers often call on, even when they have local licences on frequencies, as well as regulators who retain control of the frequencies.

2.6 Assessment of Assignment Options

Spectrum being a public resource, it is legitimate to assess what approaches best achieve public policy objectives of deployment, coverage and social policy?

We can summarise NRAs' three current basic options on assignment regimes as follows:

Table 2.1   Assignment Options

ASSIGNMENT REGIME	1	2	3
Criteria	Improved auctions with coverage obligations	Assignments prioritising investment and coverage commitments	Administrative assignment
A. Network deployment coverage	NRA discretion on coverage levels potentially high or close to 100%	Reveals operators own capabilities and vision of the market	Strong role of Government
B. Competitiveness	Residual distortion by focus on fees Oligopoly limitations Allows new entrants	Limited	Limited High focus on policy objectives
C. Frequency fee	Detrimental to investments	Presumably low in the short term	To be determined Endogenous or exogenous

In box A1 (Table 2.1), for instance, coverage obligations are defined ex-ante, with the primary focus of the assignment bid on the fee. For instance, in a recent decision, the Italian NRA, AGCom,[28] required 5G frequency users to deploy BBD or UBB networks "in all provinces" within a pre-determined timeline of 24 months for the 3600–3800 MHz band, 36 months for the 700 MHz band and 48 months for the 26 GHz band.

Assignment regime 1, *Auction with coverage obligations*, prioritises fees over investment. Scenario 2, *Assignment prioritising investment and coverage commitments* conforms to the spirit of the new EECC by prioritising industry, social and economic objectives. Assignment regime 3, *Administrative assignment* would in many countries present the risk of being considered a step backward.

It should be noted, however, that public policy objectives can change over time, based on timely assessments of market situations, with corresponding choices regarding spectrum assignment methods.

2.7 Where We Are and Future Perspectives

While the clear majority of spectrum assignments in this century have included policy-defined obligations related to coverage and deployment schedules, the financial bid has remained a strong determinant in granting the licence. The current connectivity concerns reinforce the evidence highlighted in Section 2.3 to affirm it risks delivering disappointing outcomes. There is fresh momentum to explore assignments where the fee paid to the government loses its central status and is considered but a secondary counterpart to the efficient use of spectrum resources.

The 2018 example of 4G licences renewal in France;[29] extended terms of payment introduced in countries such as Spain, Sweden and India; the possible abolition of fee maximisation in Colombia;[30] and the conditions of spectrum assignment in 2020–2021 in Denmark and Austria are indications that something is giving in the field of auctions. Japan endorsing a "comprehensive strategic approach" for 5G spectrum shows that policymakers are willing to look at new, more dynamic approaches to spectrum assignment with a stronger focus on prioritising macro-economic benefits and preserving openness for long-term competition.

Disruptive frameworks have their time: in telecommunications, the competitive paradigm has been successful in dynamising and energising the market. Auctions have allowed a competitive entry process into the newly de-monopolised telecommunications market. However, after the competitive phase, there has come a moment in which auctions have contributed to erecting entry barriers protecting strong oligopolies. Robust tools are needed to align procedures, strategies and connectivity objectives.

It would seem reasonable, especially at the time when "whatever it takes" makes auctions proceeds secondary and connectivity imperatives a priority, to manage the licensing and investment processes in an effective way.

Notes

1. R.H. Coase, October 1959. The Federal Communications Commission. *Journal of Law and Economics*, 2, pp. 1–40.
2. T. Mc Craw, 1984. *Prophets of Regulation*. Cambridge: Belknap Press.
3. Green Paper on radio spectrum policy in the context of European Community policies such as telecommunications, broadcasting, transport, and R&D/* COM/98/0596 final */
4. Federal Communications Commission Spectrum Policy Task Force, November 2002. ET Docket No. 02–135. https://sites.national academies.org/cs/groups/bpasite/documents/webpage/bpa_048826.pdf.
5. M. Cave, March 2002. *Review of Radio Spectrum Management, An Independent Review for the Department of Trade and Industry and HM Treasury.* https://web1.see.asso.fr/ICTSR1Newsletter/No004/RS%20 Management%20-%202_title-42.pdf
6. N. Smith, 24 November 2015. *Most of What You Learned in Econ 101 Is Wrong, the Theory Is Out of Date.* www.bloomberg.com/view/ articles/2015-11-24/most-of-what-you-learned-in-econ-101-is-wrong.
7. The frequent use of the term "market failure" in the telecommunications networks context is questionable. Achieving 100% penetration cannot be the universally expected outcome on all markets. "Limited market penetration" would be more appropriate.
8. https://telecoms.com/501033/french-watchdog-outlines-mid-band-spectrum-auction-rules/.
9. Ibid. Smith, *Most of What You Learned in Econ 101 Is Wrong*.
10. R.E. Backhouse and B. Cherrier, December 2017. The Age of the Applied Economist: The Transformation of Economics since the 1970s. *History of Political Economy*, 49(Supplement), pp. 1–33. https:// doi.org/10.1215/00182702-4166239, SSRN: https://ssrn.com/abstract= 2868144 or http://dx.doi.org/10.2139/ssrn.2868144.
11. T. Kuroda, B. Forero, and M. Del Pilar, June 2017. The Effects of Spectrum Allocation Mechanisms on Market Outcomes: Auctions vs Beauty Contests. *Telecommunications Policy*, 41(5–6), pp. 341–354.
12. GSMA, 22 February, 2017. *Effective Spectrum Pricing Helps Boost Mobile Services*. http://www.gsma.com/spectrum/effective-spectrum-pricing/
13. NERA Economics for GSMA, 2007. *Effective Spectrum Pricing*, p. 31. www.gsma.com/spectrum/effective-spectrum-pricing/.
14. European Commission, 4 October 2017. *Study on Spectrum Assignment in the European Union*. https://publications.europa.eu/en/publication-detail/-/publication/2388b227-a978-11e7-837e-01aa75ed71a1/language-en.

15. European Commission, *Study on Spectrum Assignment in the European Union*, p. 100.
16. C. Cambini and N. Garelli, June 2017. Spectrum Fees and Market Performance: A Quantitative Analysis. *Telecommunications Policy*, 41(5–6), pp. 355–366.
17. Thomas W. Hazlett, Roberto E. Muñoz, and Diego B. Avanzini, 2012. What Really Matters in Spectrum Allocation Design. *Northwestern Journal of Technology and Intellectual Property*, 10, p. 93. https://scholarly-commons.law.northwestern.edu/njtip/vol10/iss3/2.
18. G. Pogorel and E. Bohlin, 2017. *Spectrum 5.0: Improving Assignment Procedures to Meet Economic and Social Policy Goals, A Position Paper.* Working paper. www.researchgate.net/publication/316524026_Spectrum_50_Improving_assignment_procedures_to_meet_economic_and_social_policy_goals_A_position_paper
19. Commission Impact Assessment Guidelines (January 2009) – G. Pogorel et al., 2015. *Socio-Economic Impact of Spectrum Regulation: Competition and Consumer Protection.* ITU Spectrum Management Training Program.
20. We focus on mobile wireless and leave aside at this stage the equally critical and closely linked domain of broadcasting.
21. A. Magi, March 2017. *Assessment of the Socio-Economic Impact of Mobile Broadband Auctions.* Thesis for the Master's degree, Politecnico di Torino.
22. www.fcc.gov/auction/107/factsheet#License_Period.
23. PTS, 2011. *Open Invitation to Apply for a License to Use Radio Transmitters in the 800 MHz Band.* www.pts.se/upload/Beslut/Radio/2010/10-10534-open-invitation-800-mhz-auction-dec10.pdf.
24. https://ens.dk/ansvarsomraader/frekvenser/auktioner-og-udbud-frekvenser.
25. Telecompaper, 25 June 2018. *Danish Govt Issues Final Rules for 700 MHz, 900 MHz, and 2300 MHz Auctions*, Monday.
26. www.real-wireless.com/successful-spectrum-auction-model/.
27. For example, obligations would be contingent on government and local authorities abiding by the Broadband Cost Reduction Directive.
28. AGCom Delibera n. 231/18/CONS 74 Gliaggiudicatarideidirittid'usod ellefrequenzenellebande700MHzSDL,3600-.
29. ARCEP, 2 August 2018. www.arcep.fr/actualites/les-communiques-de-presse/detail/n/new-deal-mobile-1.html.
30. *Policy Tracker.* www.policytracker.com/colombia-may-abolish-maximised-spectrum-prices.

3

Towards a Future-Proof International Spectrum Policy?

MOHAMED EL-MOGHAZI AND JASON WHALLEY

Contents

Cellular phones will absolutely not replace local wire systems.

Marty Cooper, inventor, 1982

3.1 Introduction

One thing that is certain about the future is that we tend to act as if we are well-prepared for it, while in most cases we react to unforeseen events that we have not planned for. Similarly, the way we have planned radio spectrum at the international level has largely been reactive in character rather than proactively preparing for the future. It was the failure of Prince Henry of Prussia's message, returning across the Atlantic from a visit to the USA, to be successfully sent to President Theodore Roosevelt in 1902 that resulted in the first International Radiotelegraph Conference in 1906. And it was the sinking of the Titanic in 1912 that motivated the International Radiotelegraph Conference, held in the same year, to agree on a common wavelength

DOI: 10.1201/9781003156765-4   **47**

for the radio distress signals of ships. More recently, ITU-R World Radiocommunication Conference of 2015 (WRC-15) allocated the 1087.7–1092.3 MHz band for aeronautical mobile-satellite service to allow flights to be tracked globally in response to the loss of Malaysian Airlines flight MH370.

This is not to say, however, that there has been no preparation for future wireless developments by the international spectrum community. Preparations and studies for the almost 30 issues to be addressed at WRC-23 are ongoing, along with another 13 potential issues for WRC-27. Moreover, there have recently been several mobile allocations and International Mobile Telecommunication (IMT) identifications to meet the growing demand in data and prepare for the deployment of 4G and 5G systems.

But the main reason to focus on international spectrum policy is that we believe it is an integrated and critical element of any spectrum management debate that usually focuses on alternative approaches for managing access to radio frequencies nationally. In particular, investigating spectrum liberalisation should consider the influence of the ITU-R Radio Regulations (RR) and the different international measures related to service allocation and technology selection. The main principles of the international spectrum management regime were drafted at a time when the 'command-and-control' approach was the only way to manage spectrum in order to handle interference and to enable international operability. Yet today these principles are largely unchanged, influencing the way regulators address spectrum management and use as well as how different stakeholders (e.g., operators, broadcasters, manufacturers) interact within the national spectrum policy domain.

To ensure that international spectrum policy(s) can accommodate uncertainties regarding the future development of the wireless industry, in terms of disruptive technology or new ways of managing spectrum, it is necessary to identify those key elements that act as a remedy to abuse. These elements, which include interference protection, harmonisation, consensus, systems' spectrum identification, are Janus-faced in character – they generate benefits but can also be used to deter innovation and enforce 'traditional' ways of managing spectrum.

The rest of this chapter is divided into six main sections. In the first four of these sections, a series of options are discussed to illustrate the challenges inherent to international spectrum management and how it may change in the future. Section 3.2 uses 6 GHz to explore the choice between licensed and unlicensed and is followed by a discussion of flexibility versus harmonisation. In Section 3.4, we use IMT to explore the choice between allocation and identification, before switching our focus in Section 3.5 to the procedures used within World Radiocommunications Conferences that decide on which policy option is adopted. In the penultimate section of the chapter, we use the issue of interference to explore the tension between providing certainty on the one hand and sufficient flexibility to encourage innovation. In the final section of the chapter, conclusions are drawn.

3.2 6 GHz: Licensed versus Unlicensed

The licensed versus unlicensed debate is a major part of the spectrum management corundum, where, in the former, certainty for investment and protection against interference are major advantages. However, licence exception (i.e., unlicensed) entails non-interference non-protection spectrum usage while adhering to specific conditions set by the regulators. One argument in favour of licensing is that lack of spectrum ownership may cause market failure by discouraging incumbents to invest in technology and infrastructure (Hazlett & Leo, 2010). The alternative view is that spectrum should be regulated in a way similar to the Internet which adopts decentralised commons structure (Benkler, 1998). Hence, regulators should focus on regulating the use of equipment rather than the spectrum.

At a first glance, licensing is usually perceived as a national decision where regulators may allow license-exempt use to a particular spectrum band with different flavours (e.g., registration, light licensing). However, Article 11 of the ITU-R RR states that "no transmitting station may be established or operated by a private person or by any enterprise without a licence issued in conformity with the RR" (ITU-R, 2020c). Additionally, Article 4.4 of the RR allows the assignment of spectrum in a derogation from the RR on the condition that it does not cause or claims protection from harmful interference (ITU-R,

2020b). Therefore, in practice, the RR call for the licensing of stations while allowing for other uses on a non-interference non-protection basis. For instance, Wi-Fi devices operate on a non-protection, non-interference basis and with low powers (Anker & Lemstra, 2011). These devices usually operate in the what is called Industrial, Scientific and Medical (ISM) spectrum bands such as 2400–2500 MHz and 5725–5875 MHz (ITU-R, 2012a). These bands are designated for ISM applications and radiocommunication services operating within these bands must accept harmful interference which may be caused by these ISM applications (ITU-R, 2012a).

The recent discussion regarding the future use of the 6 GHz band sheds light on the impact of international spectrum management on a regulator's licensing decision. More specifically, there is currently a heated debate on the future use of the 5925–7125 MHz band and whether it should be used as an extension for the Wi-Fi systems in the 5 GHz or as the major mid-band for emerging 5G technologies. The issue emerged before WRC-19 where there was a proposal to discuss the identification of the whole 6 GHz band (5925–7125 MHz) for IMT as part of the agenda of WRC-23 as the band 6425–7125 GHz is already allocated to mobile service, but it is used mainly for microwave links and VSAT applications (FSS). The mobile industry perceives that sharing of 5G services with these two services is feasible (Ericsson, 2019). However, before WRC-19, the Wi-Fi lobby urged countries to oppose identifying any part of the 6 GHz band to IMT, arguing that Wi-Fi devices can co-exist with existing services unlike IMT systems which would require the relocation of these services (Wi-Fi Alliance, 2019).

During WRC-19, while the USA called for operating the whole band for unlicensed applications, China supported the opposite, namely, to utilise the whole band for 5G systems, which normally operate on a licensed basis. Eventually, it was decided to split the 6 GHz band between IMT and unlicensed application, with the upper band (6425–7125 MHz) being studied for IMT at WRC-23 (Marti, 2019). WRC-19 decided that Agenda Item 1.2 of WRC-23 is to consider identification of the frequency bands (i.e., 3300–3400 MHz, 3600–3800 MHz, 6425–7025 MHz, 7025–7125 MHz and 10.0–10.5 GHz) for IMT, including possible additional allocations to the mobile service on a primary basis, in accordance with Resolution 245 (WRC-19). Such resolution asked the ITU-R to conduct and complete for WRC-23 the appropriate studies of technical, operational and regulatory issues. The resolution also sought to conduct and complete in time for WRC-23 the sharing and compatibility studies with a view to ensuring the protection of services to which the frequency band is allocated on a primary basis without imposing any constraints on them (ITU-R, 2019f).

Following the conference, proponents of both IMT and Wi-Fi have started to strongly lobby to acquire more frequencies in the 6 GHz band, and several countries have started to take sides either by designating the whole band for Wi-Fi unlicensed operations (e.g., the USA, South Korea, Brazil), designating the lower part (e.g., UAE, South Korea), or by identifying the whole band for IMT (China) (Analysys Mason, 2019).

The argument against Wi-Fi in the 6 GHz band includes the lack of coordination between the Wi-Fi devices and incumbent services which may lead to interference. Furthermore, the different adopted frequency ranges by the different regions in the 6 GHz band will not achieve harmonisation. Designating a spectrum band for unlicensed use is usually irreversible and requires tremendous effort to re-farm it for other uses (GSMA Intelligence, 2020). However, IMT identification implies licensing the spectrum to mobile operators on an exclusive basis to enable economies of scale and provide the necessary certainty for investment. One other argument in favour of licensed spectrum is that operators must have control over the spectrum to enable the required data rate anytime anywhere and to provide sufficient capacity for vertical users. In addition, sharing with the incumbent is regarded as being easier and could be treated on a case-by-case basis as the positions of the transmitters are known and coordination measures are included in the licence conditions. Interestingly, one claim that has been made is that Wi-Fi loading is declining due to the better user experience from 4G and 5G unless the Wi-Fi utilises a reliable Fibre to the Home (FTTH) connection. This is also supported by a tendency towards providing unlimited data plans (Coleago Consulting, 2020). This is, of course, in contrast to Cisco internet reports that indicate offloading will increase significantly with the introduction of 5G (Cisco, 2020).

The importance of the 6 GHz band for Wi-Fi is highlighted by the fact that Wi-Fi 6 is different from previous generations of Wi-Fi in terms that it provides broader spectrum bandwidth. This will meet the increasing amount of offloaded traffic from 5G as almost 80% of all mobile broadband network is expected to be offloaded to Wi-Fi by 2025 (3GPP, 2019). Furthermore, Wi-Fi 6 could operate in a bandwidth of up to 160 MHz and provides speeds of up to 9.6 Gbps, which is quite similar to 5G speeds. Additionally, being adjacent to 5 GHz, already utilised for Wi-Fi, would facilitate the manufacture of these new Wi-Fi devices due to similarities between the designs of the two radio bands (Ofcom, 2020).

It has been suggested that between 500 MHz and 1 GHz of new spectrum will be needed in 2025 to meet increasing Wi-Fi demands (Jones, 2017). This is supported by the fact that the use of Wi-Fi has increased by over 50% during Covid-19 (Bhatia, 2020). Another argument in favour of Wi-Fi is that it decreases the investment needed to establish a ubiquitous 5G network in order to make the service affordable to the users (DSA, 2021).

An argument against IMT is that sharing studies in previous WRCs have shown that sharing is not possible between IMT and existing services (Royabalat, 2020). Therefore, even if the IMT identification is approved, it would be with additional constraints that may not be beneficial for the deployment of 5G in the same band (Royabalat, 2020). In particular, sharing conditions for IMT in the 6 GHz band may be restricted due to the existence of incumbent services such as earth exploration-satellite (EESS) (passive) and space research (SRS) (passive) services in the 6425–7075 MHz and 7075–7250 MHz bands (SFCG, 2020).

Another issue with the IMT identification in the 6425–7025 MHz band is that it has been discussed only within Region 1. Thus, even if WRC-23 approves the identification, it will not achieve global harmonisation. Instead, it will disturb the global deployment of Wi-Fi 6 in the 6 GHz band. However, as Wi-Fi and IMT can operate under the existing co-primary mobile allocation in the 6 GHz band, the operations of current services should not be negatively impacted.

While we are not in a position to argue for or against IMT or Wi-Fi deployments in the 6 GHz band, the issue highlights a number of important features of the impact of international spectrum management on licensing. More specifically, it is clear that IMT needs spectrum identification from the ITU-R to achieve economies of scale and harmonisation of use around the world. In contrast, unlicensed use usually requires nothing from ITU-R and needs no identification in the RR as it operates on non-interference non-protection basis.

In fact, with the exception of WRC-2003, which decided to allocate the bands 5150–5250 MHz and 5470–5725 MHz on a primary basis to the mobile service for the implementation of wireless access systems (WAS), including Radio Local Area Networks (RLANs) (ITU-R, 2003), there has never been a case where the ITU-R designated a spectrum band for the operation of Wi-Fi. The main objective at that time was to give these systems an appropriate ITU allocation status and to find an alternative band(s) instead of the ISM allocation in the 2.4 GHz and 5 GHz bands (CEPT, 2000). In 2003, the ITU-R designation of a primary mobile allocation was needed to harmonise WLAN frequency use globally and to legitimise unlicensed WLAN operations in developing countries that are not familiar with the concept of unlicensed services (Sung, 2003). Even with such a primary allocation, WLAN devices operate on a secondary basis while utilising techniques such as Dynamic Frequency Selection (DFS) not to cause interference with radar systems that also have a primary allocation (Arefi & Cordeiro, 2020).

This leads to the dilemma for some countries that if it is not in the RR, then it is not allowed, and, for other countries, if it is not in the RR, it is not prohibited. For the case of Wi-Fi in the 6 GHz band, it seems that at some point there was no need for international action. For instance, almost two years before WRC-19, the EU issued a mandate to CEPT to study and identify harmonised technical conditions for WAS/RLANs in the 5925–6425 MHz band to provide wireless broadband services (Ofcom, 2020). CEPT published several studies examining the coexistence between the incumbents and RLAN systems in the 5925–6425 MHz and adjacent bands (ECC, 2019). Similarly, the Federal Communications Commission (FCC) published in 2017 a Notice of Inquiry seeking input on 'flexible access' in several bands, including 5925–6425 MHz and 6425–7125 MHz

(Ofcom, 2020). Unlicensed use of Wi-Fi was of particular interest to the FCC.

Recently, countries have started to take clear decisions regarding Wi-Fi. In November 2020, CEPT approved a decision on the harmonised use of the frequency 5945–6425 MHz band for WAS, including RLAN (ECC, 2020). The decision allows low power indoor (LPI) use that may operate both indoor and outdoor. In April 2020, the FCC decided to make the 5.925–7.125 GHz band available for unlicensed use where indoor low-power operations can operate over the full 1200 MHz and standard-power devices operate in 850 MHz in the 6 GHz band. The coordination between the incumbent services and Wi-Fi devices is determined to be through an automated frequency coordination (FCC, 2020). The UAE has identified in December 2020 the 5925–6425 MHz band to Wi-Fi for indoor use (Emirates News Agency, 2020). However, KSA has released the full 6 GHz band (1200 MHz) to Wi-Fi in March 2021 (Sbeglia, 2021).

There is no harmonisation for the use of Wi-Fi in the 6 GHz, even within Region 1. Neighbouring countries such as UAE and KSA have different spectrum operations for devices that are expected to be sold on a large scale and move across borders. While countries are sovereign regarding such decisions, it seems that the IMT discussions at WRC-23 in the upper 6 GHz may have an influence on the choice of Wi-Fi operating band even if the operation of the latter is not identified within the RR. Wi-Fi proponents would have benefited from a similar mobile primary allocation in the 6 GHz band in addition to the 5 GHz decision that was taken by WRC-03. In such a case, the harmonisation would have been facilitated for Wi-Fi devices and any decision regarding IMT would recognise such an allocation. Instead, there is now a struggle between IMT and Wi-Fi in the 6 GHz band, with one way out possibly being the introduction of the concept of tuning range where IMT devices are capable to operate in a range of frequencies where the actual operation is conditioned by the licensed and approved frequencies within each country (Bhatia, 2020).

In general, the idea of having unlicensed devices operating on a non-interference non-protection basis without any recognition from the ITU-R should be reconsidered as the benefits of regional or global harmonisation through the RR could exceed the difficulties associated with acquiring such recognition. The 6 GHz issue has highlighted the confrontation between China, as one of the leaders in the 5G era, and the USA, the main advocate of unlicensed use. From another perspective, the issue may be perceived as the face-off between supporters of innovation and liberalisation in spectrum use, as emphasised by the case of Wi-Fi and their supporters, mainly from the OTT and Internet giants based in the USA, and the traditional advocates of exclusivity and governmental control over wireless activities, which is the Chinese 5G sector that is extensively supported by its government.

Having said this, it can be argued that the USA should re-consider its international campaign strategies given that innovation in spectrum use would make use of a binding decision by the ITU. This requires lobbying not only within CITEL but also in Regions 1 and 3, and it would be years before such innovation could be practically deployed. This was exactly what China did to promote the concept of IMT in the 6 GHz band. In particular, China managed to obtain a study of most of the upper 6 GHz band (6450–7025 MHz) within Region 1 even though the same issue was opposed in Region 3 due to the opposition from other Asian countries within APT (The Asia-Pacific grouping within the ITU). This was mainly due to pressure from the Chinese mobile industry, supported by their government, and the demands of the African countries, which originally called for studying the whole 6 GHz band for IMT (Paoletta, 2019). It is worth noting that some African countries heavily rely on Huawei equipment (Wallsten, 2020).

3.3 UHF: Flexibility versus Harmonisation

One of the most controversial issues at the last WRC was the future of the ultra-high frequency (UHF) band and the possibility of additional mobile allocation in Region 1 under AI 1.5 of WRC-23. This reflects the required balance between harmonisation, which is at the heart of international spectrum management, and flexibility that enables countries to diverge from the RR regional or global harmonisation. In order to address the issue, it is important to understand that historically most of the UHF band (470–862 MHz) was planned for analogue terrestrial broadcasting services in Region 1. In

2006, the Regional Radiocommunication Conference 2006 (RRC-06) planned the digital terrestrial broadcasting service in Region 1 and in the Islamic Republic of Iran to be in the 174–230 MHz and 470–862 MHz frequency bands (ITU, 2006). Shortly after wards, WRC-07 approved an additional allocation in the 790–862 MHz band to mobile service (ITU-R, 2007a). WRC-12 added an additional mobile allocation and IMT identification in the 698–790 MHz band in Region 1 (ITU-R, 2012d).

During WRC-15, the USA proposed adding an allocation to the mobile services and identification for IMT in the 470–694/698 MHz range except for the 608–614 MHz band in Region 2. The proposal sought to achieve a global harmonised allocations for mobile service in the UHF band to ensure flexibility for countries, enabling them to provide either broadcasting or mobile service according to their needs (USA, 2015). Similarly, four Arab countries in Region 1 proposed at the beginning of the conference a footnote to identify the band for IMT in the 470–698 MHz band in those countries wishing to, noting that the identification is subject to the GE-06 agreement that protects broadcasting service (Egypt et al., 2015).

However, the majority of the regional groups opposed the mobile allocation in the 470–694 MHz band, arguing that was extensively used by broadcasting services, and that large separation distances are needed between mobile and broadcasting services (ASMG, 2015; ATU, 2015; CEPT, 2015; RCC, 2015). Eventually, WRC-15 agreed to have a new AI for WRC-23 to review the spectrum use and needs of existing services in the 470–960 MHz frequency band in Region 1 (ITU-R, 2015e).

In Region 2, only the Bahamas, Barbados, Canada, the USA and Mexico were able to obtain identification in the 470–698 MHz frequency band, albeit with caveats regarding the broadcasting service, whereas Belize and Columbia obtained identification only in the band 614–698 MHz. In Region 3, a few island countries (Micronesia, the Solomon Islands, Tuvalu and Vanuatu) obtained identification in the 470–698 MHz band, whereas Bangladesh, Maldives and New Zealand succeeded in identifying the 610–698 MHz frequency band (ITU-R, 2015a). Furthermore, many countries failed to add their names into the footnote as their neighbours objected (e.g., Pakistan,

Papua New Guinea, India), whereas many others welcomed the footnote on the condition that their neighbours did not add their name to it (ITU-R, 2015d).

Although it was already agreed at WRC-15 to have an agenda item in WRC-23 on the UHF band, WRC-19 did, in theory, have the capacity to review that agenda item and to suppress it if this was agreed by the participants. The ATU position during WRC-19 was to pay extra attention to the proposed agenda item given that the majority of African countries plan to extensively use the 470–694 MHz band for broadcasting, while CEPT presented no modification to the item. In contrast, ASMG requested mentioning the possibility for IMT identification explicitly, while RCC proposed to suppress the agenda item.

Due to the fact that the agenda item was the result of a (delicate) compromise at WRC-15, WRC-19 eventually confirmed and agreed to have an agenda item for WRC-23 without any modification to the potential agenda item proposed at WRC-15. Agenda item 1.5 is to review the spectrum use and needs of existing services in the 470–960 MHz frequency band in Region 1 and consider possible regulatory actions in the 470–694 MHz frequency band in accordance with Resolution 235 (WRC-15). Within the context of this agenda item, the ITU-R would review spectrum use and study the requirements of existing services (broadcasting and mobile except aeronautical mobile) within the 470–960 MHz frequency band in Region 1. The ITU-R will also conduct appropriate sharing and compatibility studies.

Following WRC-19, CPM23-1 established a new Task Group 6/1 (TG 6/1) to deal with WRC-23 agenda item 1.5 as part of Study Group 6, which is responsible for broadcasting services (ITU-R, 2019a). It was also decided that Working Party 6A is to undertake a review of the spectrum use and study the spectrum needs of the broadcasting service, taking into account the GE06 Agreement, within the 470–960 MHz frequency band in Region 1. In addition, the relevant Working Parties of Study Group 5 were also to review the spectrum use and needs of mobile (except aeronautical mobile) services within the same frequency band.

The UHF issue clearly highlighted the struggle between proponents of harmonisation, who called exclusive allocation of the band to broadcasting, and the advocates of flexibility, who pleaded for the introduction of more flexibility into the band. Meanwhile, those individual countries that have been seeking flexibility have, in most cases, not been able to achieve this. For example, during WRC-15, eight Arab countries, with a combined population of over 200 million, were not able to introduce IMT identification in the 614–698 MHz band in alignment with Region 2 countries (El-Moghazi, 2017).

Another example is Finland, which was in favour of extending the mobile allocation and IMT identification to the 470–698 MHz band unlike all other CEPT countries (Finland, 2015). Moreover, many Latin American countries were not able to add themselves to the Region 2 footnote in the 470–698 MHz band. Even Columbia, which just about added itself to the footnote in Region 2 in the band 614–698 MHz, made a statement during the conference plenary that it will not implement IMT systems, even after having the identification, until it reaches a coordination agreement with Brazil and Ecuador (ITU-R, 2015d). Similarly, India, which is a large country, indicated several times that the application of IMT in its territory in the 614–698 MHz band would not cause interference to its neighbours' broadcasting services. However, such request was eventually declined by its neighbours (Youell, 2015).

In essence, the RR accommodate primary mobile and broadcasting allocation in most of the UHF band (470–862 MHz) in Region 3. There has been even a recommendation that establishes protection for land mobile systems from terrestrial digital video and audio broadcasting systems in the VHF (174–230 MHz) and UHF (470–862 MHz) shared bands (ITU-R, 2006). For the case of the adjacent operation of mobile and broadcasting services, the ITU also set the unwanted emissions from the former to the latter. It is understandable that too stringent limits may lead to an increase in size, cost or in complexity of IMT radio equipment (ITU-R, 2015b). Therefore, it may seem unreasonable to oppose flexibility between mobile and broadcasting services in Region 1 as it has been acceptable, at least in the other ITU regions, to have co-primary allocation between broadcasting and mobile.

It is understood that Region 1 is different in terms of having a regional agreement such as GE-06 that governs the use of broadcasting services. Altering such an agreement requires another regional conference, and even WRC does not have the authority to change such a plan. While the RR could be altered through a WRC agenda item, the GE-06 agreement can only be modified by another RRC. In addition, there are currently several countries that are largely dependent on traditional terrestrial TV (e.g., Italy in Europe and South Africa in Africa), and their deployments have an influence on their neighbours due to the propagation characteristics of the UHF band.

Another dilemma with RRC-06 is that although Iran is part of the plan due to its proximity to many countries in Region 1, it is not a member of this region. Furthermore, Iran largely relies on terrestrial broadcasting across most of the UHF band (470–862 MHz) to provide television services. The propagation characteristics of the shared borders that Iran has with countries in Region 1, coordination is relatively difficult (ITU-R, 2014a). Of course, co-sharing between broadcasting and IMT, in particular, as an application of mobile service, may be difficult and require large separation distances, especially in the summer (ITU-R, 2017a). However, it is also recognised that there are some measures that could decrease the possibility of interference such as reducing IMT antenna height or tilting it downwards (ITU-R, 2014b).

UHF is arguably caught between broadcasting and mobile industries. From the broadcaster's perspectives, video consumption will remain mostly in-home, delivered via broadcasting and increasingly through streaming. Providing broadcasting services via 5G will not replace traditional broadcasting platform due to their incompatibility with GE-06 and the channelisation differences (5 MHz for 5G and 8 MHz for GE-06) (Hemingway, 2020). The UHF band has witnessed the continuous encroachment from the mobile industry, with the 900 MHz, 800 MHz and 700 MHz bands being identified for IMT in, respectively, WRC-2000, WRC-07 and WRC-12 (Kholod, 2020). Following these IMT identifications, the remaining broadcasting band in the 470–694 MHz band is just 57% of that planned in 2006. In Europe, it is forecasted that linear viewing will remain the main way of viewing TV content for the foreseeable future, while time-shifted and on-demand (non-linear) TV will continue to grow (ECC, 2014). However, in the USA, the FCC utilised an incentive auction where service flexibility is provided in practice between the broadcasting and mobile services by offering incentives to existing users to return unused or unneeded spectrum. Those users who returned spectrum on a voluntary basis received financial compensation. The auction cleared the spectrum in the 600 MHz band between the 614–698 MHz of broadcasting services and resulted in $19.8 billion being spent (ITU-R, 2019c).

To this end, it is recognised that while the EU decision to designate the 470–694 MHz frequency band for the use of terrestrial provision of broadcasting services, including free television, and for use by wireless audio PMSE at least until 2030 (EU, 2017), such use is to be reviewed by 2025 to assess technology and market development (Lamy, 2014). Therefore, while it may be quite difficult to reach a decision on the future use of the (470–698 MHz) band, one possibility is to postpone the discussion on the UHF band to WRC-27 in order to allow for more time to review the situation in the light of the rapid changes in user behaviour. While there is the possibility of agreeing to not change the item, there is also the risk of restricting the use of the band for almost a decade considering that including an agenda item in WRC and then studying it could take eight years. In such a case, and assuming that there is a decrease in the linear TV penetration following WRC-23, countries may start to use the band for mobile and IMT without having allocation or identification in the RR similar to the case of C-band in Europe (3.6–3.8 GHz) (El-Moghazi & Whalley, 2019). Otherwise, countries may start deploying IMT in the 470–698 MHz without waiting for a decision to be reached at WRC-23. Such an unliteral approach was adopted by the USA regarding the UHF band before WRC-15, enabling it to achieve faster UHF spectrum re-farming as it did not have to wait for a decision from the ITU (Frieden, 2019a).

3.4 IMT: Allocation versus Identification

In this section, we examine the concepts of service allocation and system identification considering the cellular mobile systems that require mobile service allocation and IMT identification in the RR. More specifically, over the course of the last years, several systems, such as

HAPS and IMT, have subsequently acquired identification in several bands that may be considered as a barrier to emerging technologies. Furthermore, a number of technologies and systems (e.g., CRS, ITS, RSTT) were studied by the ITU-R, with no identification being agreed.

In general, allocating spectrum to the different services is the main responsibility of the ITU-R. The international table of frequency allocation, as incorporated within Article 5 of the RR, divides the frequency band from 9 kHz to 400 GHz into smaller bands that are then allocated to more than 40 radiocommunication services (ITU-R, 2001). Dividing the spectrum according to the type of service and global harmonisation of spectrum allocations is the ITU historical methods of mitigating harmful interference.

It is not usual for the RR to identify spectrum for specific uses except for a few systems or technologies. In the early stages of the development of IMT, there was a debate at WARC-92 on whether to allocate or identify spectrum for Future Public Land Mobile Telecommunication System (FPLMTS), the former name for IMT-2020 (U.S. Congress Office of Technology Assessment, 1993). It was agreed that identification was to be used on the condition that it does not preclude other uses of the spectrum. The IMT standardisation process has been associated with identifying the radio spectrum bands to be used by these IMT standards (e.g., 2 GHz band). The first step was at WARC-92 when the conference identified the 1885–2025 MHz and 2110–2200 MHz bands for countries wishing to implement FPLMTS (ITU-R, 2016). Since then, several other bands have been identified for IMT, such as 900 MHz, 1800 MHz and 2100 MHz.

Although all IMT systems are expected to operate within bands identified for IMT, there have been several cases where they have operated in bands allocated to mobile services (El-Moghazi & Whalley, 2019). This includes the L-Band (1.452–1.492 MHz), which has been used by CEPT countries without alignment with IMT identification at WRC-15, the 26 GHz band (24.25–27.5 GHz) that was decided to be used for IMT before the WRC-19 decision on the band and the 28 GHz band (27.5–29.5 GHz) that has been used by the USA for IMT without any identification within the RR. There has also been a case where CEPT countries and several Arab countries have used IMT systems in bands not allocated to mobile service nor identified to IMT (3.6–3.8 GHz).

The last case motivated WRC-19 to include an agenda item in WRC-23 to consider primary allocation of the 3600–3800 MHz band to mobile service within Region 1 and take appropriate regulatory actions in accordance with Resolution 246 (WRC-19). This resolution invites ITU-R to conduct sharing and compatibility studies in time for WRC-23 between mobile and other services allocated on a primary basis within the 3600–3800 MHz frequency band and adjacent bands in Region 1. This would ensure protection of those services to which the frequency band is allocated on a primary basis and not impose undue constraints on the existing services and their future development.

Other example of system identification is for high-altitude platform stations (HAPS), which provide fixed broadband connectivity that would enable wireless broadband deployment in remote areas with minimal ground-based infrastructure (ITU-R, 2015c). Although HAPS have identification in a number of spectrum bands, several of these bands have not been fully utilised in the past due to particular physical constraints and technical and regulatory conditions (ITU-R, 2018b). Similarly, the IMT identification in the C-band, which was decided in WRC-07, was not fully utilised due to the stringent conditions in favour of FSS.

There are other cases where some sort of identification was discussed but was not agreed by the ITU-R. For instance, the regulatory measures that could enable deployment of Cognitive radio system (CRS) was discussed at WRC-12 (ITU-R, 2007b). The issue was raised during WRC-07, with Resolution 956 inviting the ITU-R to study measures such as the need for a database that can assist in the determination of local spectrum usage (ITU-R, 2007d). Eventually, WRC-12 did not decide on any particular measure with regard to CRS, and no spectrum was allocated to CPC as it was recognised that CRS are technologies and not radiocommunication services. It was also agreed that the examination of the implementation and use of CRS in radiocommunication services should continue without the need for consideration in the next WRC (ITU-R, 2012b). WRC-12 also recommended that any radio system implementing CRS technology should operate in accordance with the provisions of the radio regulations and that the use of CRS does not exempt administrators from their obligations in accordance with the RR (ITU-R, 2012c).

Short range devices (SRD) were also discussed during WRC-12 under AI 1.22. This agenda item examined the effect of emissions from SRDs on radiocommunication services, with further studies being required regarding the emissions from SRD in the frequency bands designated in the RR for ISM applications (ITU-R, 2007c). While having spectrum identified within the RR for software defined radio (SDR) may look appealing, to encourage economies of scales and harmonisation, the studies at that time concluded that there is no need to modify the RR or to continue the studies of SRD harmonisation (Al-Rashedi, 2014). Instead, the ITU-R Recommendation SM.1896 includes the frequency ranges for global or regional harmonisation of SRD, indicating the ranges for the potential global harmonisation of SRD.

The issue of spectrum identification for railway radiocommunication systems between train and trackside (RSTT) was discussed during WRC-19, with two views emerging. The first was to harmonise spectrum through a resolution from WRC-19 in order to provide a stable radio regulatory environment for the railway industry and to minimise risk of interference, especially for cross-border operations. Alternatively, European countries argued that they have already a framework for the operation of RSTT through GSM-R and that there was no need for such an identification (SFCG, 2020). Similarly, Agenda Item 1.12 of WRC-19 discussed spectrum identification for Intelligent Transport Systems (ITS). The APT countries, including Japan, had a similar position with regard to ITS: they supported the consideration of the 5850–5925 MHz frequency band as a global harmonised frequency band for ITS. Meanwhile, CEPT felt that the existing regional harmonisation measures for ITS in the 5855–5925 MHz band were sufficient with the consequence that no changes to the RR were required (SFCG, 2020).

Regarding Massive machine-type communication (MTC) systems, WRC-19 addressed the possible harmonised use of spectrum to support the implementation of narrowband and broadband machine-type communication infrastructures. Studies prior to the conference concluded that there was no need for any regulatory action within the RR with regard to the spectrum intended for use by MTC applications and that instead the harmonised use of spectrum to support these developments could be facilitated through ITU-R Recommendations or Reports. In essence, MTC applications can operate within identified spectrum for IMT without an explicit list of harmonised spectrum bands. This includes the 733–736 MHz and 788–791 MHz bands, which are considered by some countries in Region 1 as a potential option for harmonised use of IMT-based narrowband MTC networks deployment (ITU-R, 2018d).

While the previous cases highlighted several systems that do not have identification in the RR but are, instead, included in ITU-R reports or recommendations that do not have treaty status, these are other cases where a form of recognition in the RR has occurred. For instance, WRC resolution 646 encourages administrations to use, as much as possible, harmonised frequency ranges for public protection and disaster relief (PPDR) while taking into account the national and regional requirements. Similarly, Recommendations ITU-R M.2121–0 (01/2019) asks that administrations should consider using the 5850–5925 MHz frequency band for current and future ITS applications. The difference between ITS and PPDR is that the frequency bands in the former are mentioned in an ITU-R recommendation, which is less binding than the case of PPDR where the frequencies are mentioned in an ITU-R resolution.

There are other systems that have recognition in the RR with some sort of indirect identification. For instance, while Multiple Gigabit Wireless Systems (MGWS) are recognised in ITU-R recommendation M.2003–2 to benefit from harmonised frequencies in the 60 GHz band (ITU-R, 2018a), MGWS are also recognised indirectly in Resolution 241 (from WRC-19). This is part of the RR and resolves that

> administrations wishing to implement IMT in the frequency band 66–71 GHz, identified for IMT, which also wish to implement other applications of the mobile service, including other wireless access systems in the same frequency band, consider coexistence between IMT and these applications.
>
> (ITU-R, 2019e)

It is implicitly understood that these other wireless systems could include MGWS even if the 66–71 GHz band is formally identified but MGWS can operate within the mobile allocation without an identification. In fact, MGWS supporters attempted to include more explicit recognition in the WRC-19 resolution with the USA calling for equal access to the band by MGWS and IMT because the former was recognised by the ITU-R in Recommendation M. 2003–2 to potentially operate globally in the 60 GHz band (ITU-R, 2018a). The USA, however, encountered opposition because the WRC-19 agenda included only the discussion of identifying the 66–71 GHz band to IMT. The identification of spectrum bands for MGWS would have been outside the conference mandate.

Another interesting observation is that there has been a form of identification within the already identified spectrum. Agenda Item 1.4 of WRC-23 is to consider the use of HAPS as IMT base stations (HIBS) in the mobile service in certain frequency bands below 2.7 GHz that have already been identified (regionally or globally) for IMT. In other words, it considers having an additional layer of identification for HIBS within the layer of traditional IMT identification in bands allocated to mobile service. Such identification involves protecting existing services without imposing any additional technical or regulatory constraints on them.

It is clear, therefore, that identification is important in that it sends strong signals to the manufacturers of equipment that these are the bands that they should focus on when it comes to developing mobile technology in order to achieve harmonisation, roaming and economies of scale. Identification also sends a signal to the regulators to ensure that they have mobile allocations that could be used for IMT at an appropriate time subsequent to when they decide to introduce it into their countries. Furthermore, the harmonisation of identified spectrum has several other benefits, including efficient planning and border coordination, the development of compatible networks and effective services and equipment interoperability (ITU-R, 2017b, 2018c).

However, identification may be perceived to limit the spectrum available for services (e.g., cellular mobile phones) that have only been identified to IMT rather than being allocated to mobile radiocommunication system. Hence, as supply decreases, prices for the spectrum would increase as illustrated by the 3G auctions in some European countries. At that time of these auctions, the supply of spectrum was limited to the 2 GHz band for IMT-2000, and operators were concerned that missing out on this spectrum would influence their chances to provide 3G services. Following that, IMT spectrum extended to other frequency bands but regardless the idea of identification makes the spectrum more valuable.

The issue with IMT identification is that it sends signals that this spectrum will be used only for IMT regardless of the other services that have primary allocations in the band. This may explain the huge resistance of the broadcasting and satellite industries to IMT identification in the UHF and C-Band, respectively. Indeed, there are cases where it is clear that there is no need for spectrum identification such as that of CRS, where it is recognised that any system of a radiocommunication service that uses CRS technology within a given frequency band will do so in accordance with the RR governing the use of that band (ITU-R, 2014c). However, identification could also provide less advantage to those systems that do not have spectrum identification compared to IMT even if they can operate within the allocated radiocommunication service such as mobile. That was clear in the case of the WiMAX where its proponents sought to be part of the IMT family to access IMT identified spectrum. Considering that not all new innovative systems have the capacity to lobby through the ITU-R to acquire spectrum identification, this is arguably a barrier for emerging technologies. For instance, the call for identification for systems such as HAPS and HIBS is mainly driven by prominent groups of companies. For instance, HIBS is mainly promoted by Softbank from Japan, where HAPS are supported by the European aviation industry (e.g., Thales, Airbus) (Paoletta, 2019). However, SDR systems operate through ITU-R recommendation and could not achieve identification because there was no strong (united) industry supporting the harmonisation of frequencies for them.

The two cases of RSTT and ITS reveal how countries have different perceptions of the need for spectrum identification. In both cases, European countries, which have a robust institutional political and economic framework, appear not to need global identification as they have already regional designations. For instance, CEPT designated parts of the 5855–5925 MHz band in 2008 for use by ITS (ITU-R, 2018d). This is in contrast to a country like Japan where RSTT is of a great importance and which sought harmonisation for the frequencies for these systems.

One observation regarding having systems such as MTC that can operate under IMT identification is that it entails that these systems would probably operate on a licensed basis through the mobile operators while utilising 3GPP standards that are part of the IMT family. It does appear that having spectrum identification of IMT empowers the standardisation activities of 3GPP MTC standards. Even so, there is a considerable uncertainty associated with unlicensed spectrum, which is typically not identified for IMT, and is utilised via proprietary MTC standards. The issue can be mitigated if those proprietary technologies result in a single standard that operates on unlicensed basis (Webb, 2018).

Having said that, identification comes with a price that may be expensive for some cases. To illustrate that, the case of the potential HIBS identification in bands below than 2.7 GHz is informative. Such identification requires conducting sharing and compatibility studies to ensure the protection of services in the frequency band which have been allocated on a primary and adjacent services. This include the protection of several passive services with strict operation requirements (e.g., EESS and SRS in the band 2690–2700 MHz) (SFCG, 2020). Therefore, although HIBS acts as an extension of existing IMT coverage by providing extra base stations that can support larger coverage, they may have identification with more strict conditions than traditional IMT systems operating in the same band. In addition, one must acknowledge that there are elements of flexibility inherited in bands where there is no spectrum identification for a particular system or technology. This includes the size of spectrum bandwidth needed to meet particular national requirements and the conditions of usage similar to the case of PPDR (ITU-R, 2019g). Identification may not also result in better utilisation of the spectrum as in the case of HAPS where several bands were identified for years without widespread implementation.

To this end, it is argued that being in the RR is the last choice of ITU member states because as it is a treaty and changing it is quite difficult unlike the case of ITU-R reports and recommendations. For cellular mobile services, the level of certainty and the size of the industry requires identification within the RR. In addition, it is widely accepted that IMT technologies, with their spectrum identification, allow other applications to operate within such an identification. This include PPDR, MTC/IoT/M2M, ITS (ITU-R, 2019b). For instance, the frequency arrangements A8 and A9 allow bandwidths of 5 MHz and 3 MHz, respectively, which is not suitable for typical IMT applications as they require larger bandwidth.

3.5 WRC: Consensus versus Voting

The World Radiocommunication Conference (WRC) is held every three or four years to revise the ITU-R RR, which is the international treaty governing the use of radio-frequency spectrum and the geostationary-satellite and non-geostationary-satellite orbits. The conference also addresses any radiocommunication matter that is of a worldwide character. Within the ITU-R, voting is rarely used at WRCs, and decisions are instead reached based on consensus even if they are against the interests of some as a way of compromise. Consensus is understood as "the practice of adopting decisions by general agreement in the absence of any formal objection and without a vote" (ITU-R, 2019d).

Regional groups have a tremendous impact on WRC discussions. In particular, the USA has a great influence on the discussions within Region 2 if it manages to persuade the other CITEL member countries to support its position. This would mean that almost all of ITU Region 2 countries, with the exception of Venezuela and Cuba, would be supporting its position. Europe has a larger block of countries within CEPT, which includes 48 countries. However, CEPT countries need to coordinate with the bigger block of African and Arab countries within Region 1. Russia, a member of CEPT, is also the leader of the RCC countries, and it uses that voting power to have a bargaining power through blocking and intervening in matters that may even not be related to Region 1.

However, regional organisations usually adopt a top–down approach where countries trade-off positions on unrelated issues and tend to reach a 'package deal' (World Radiocommunication Report, 2015). Moreover, while the concept of consensus within international spectrum management is related to the importance of spectrum harmonisation, for the wireless telecommunication industry, where adopting an unliteral approach would not achieve economies of scale and roaming of devices is critical, the concept could be considered as a double-edged sword due to the ability of a handful of countries to abuse the procedures and block the entire discussion. Consensus has also the disadvantage of countries tending to negotiate and trade-off positions on unrelated agenda items that are discussed simultaneously during WRCs. For instance, the USA approved the IMT identification in the 66–71 GHz band dedicated for unlicensed operation in order to achieve a consensus on less restrictive conditions for IMT identification at the 24 GHz band during WRC-19 (Paoletta, 2019).

We have also previously highlighted how the discussion related to the identification of IMT in the 470–698 MHz band during WRC-15 demonstrated the influence of these procedures on the sovereignty of individual countries regarding spectrum use in their territories (El-Moghazi, 2017). More specifically, several countries, such as Finland, Egypt and India, were unable to obtain the identification due to objections from their neighbours even if they were willing to commit not to cause interference. Moreover, the position of Iran, a country in Region 3, regarding additional mobile allocation in the UHF did have a restrictive influence on several countries in Region 1. Iran is part of the GE-06 plan that covers itself and Region 1 countries (ITU-R, 2015d). Regardless of the context and that several countries oppose co-primary mobile allocation in the UHF, the main problem is that countries use their collective numbers to decide what other countries will deploy in their spectrum. This seems to be at odds with practices elsewhere within the UN system.

In addition, the protracted ITU-R decision-making procedures have raised a lot of concerns in recent years, especially with regards to mobile services. For instance, the identification of 3.4–3.6 GHz for IMT in Region 1 took eight years to be decided at WRC-15 following the decision of WRC-07 to identify the band (Youell, 2015). In the 5G era, the time lag between the development of technologies and ITU-R decisions has been getting longer, forcing several countries to decide on 5G development in bands that were to be considered at WRC-19 for IMT before the conference (El-Moghazi & Whalley, 2019). In the past, extraordinary WRC conferences for specific radio-communication services have been held, and there is a compelling argument to revise the timeframe of WRC, in general, to accommodate some issues reflecting their rapid development while also recognising the danger of reaching the wrong decision due to the lack of relevant studies.

Another issue that is worth highlighting is the limited involvement of the public in ITU-R decision-making procedures given that the RR are treaty documents. To understand the issue, it is worth highlighting the discussions at another ITU conference, World Conference on International Telecommunications (WCIT), which was held in December 2012 to review the International Telecommunication Regulations (ITRs) (ITU, 2012). The ITRs are another global treaty document, which were firstly adopted at the 1988 World Administrative Telegraph and Telephone Conference (WATTC-88) (Industry Canada, 2011). The ITRs regulate international telecommunications services in several areas, including traffic flow, quality of services and international routing (Internet Society, 2012). The proposed changes to the ITRs before WCIT included moving oversight of parts of the Internet from the non-governmental multi-stakeholder mechanisms such as ICANN to the ITU (Gross & Lucarelli, 2011).

The WCIT draw a lot of attention and there were calls for greater public involvement. WRCs are similar to WCIT in that they both shape the future of global connectivity, and, accordingly, it is important for the public (i.e., end users) to be involved in the decision-making process, at least within study groups. The ITU-R should encourage individual experts and NGO to at the very least submit their views for consideration without being an academic member of the ITU. The model of decision-making procedures within the ITU-R can arguably be described as 'technocratic', where the discussion and decision-making process is limited to few experts in the field which limits transparency and undermines public trust (Weyl, 2020). There has been a call for a more inclusive multi-stakeholder model to be adopted by ITU. This would include actors other than the traditional governmental delegates that reflect how technologies such as 5G have developed (Frieden, 2019b). There is, therefore, a need to involve new stakeholders in the telecommunications industry including OTT companies (e.g., Facebook) and leading Internet companies (e.g., Google) that are increasingly involved in wireless telecommunication issues in relevant discussions.

Another important lesson from WCIT is that before the conference voting was perceived not to be an option at all, with all decisions being reached via consensus. Ultimately, voting was used, and many countries did not sign the final agreement (Downes, 2012). Although voting has not been used within WRC since 1995, there have been a few occasions where it nearly happened (El-Moghazi, 2017). In general, voting was used at WCIT when there was a big divide between the proponents of altering the structure and openness of the Internet and those countries that wanted to keep the ITU away from the Internet (Downes, 2012). The formalisation of such a sharp difference of opinion has largely been avoided at WRC due to the flexibility inherent to its rules and regulations, with Article 4.4 of the RR, in particular, allowing deviation from the treaty as long as it does not cause interference or protection is claimed. Countries like the USA were able to implement IMT without formal spectrum identification in the RR (e.g., 28 GHz).

Another incident highlights the need to revise the ITU-R decision-making procedures. Following WRC-19, Iran submitted to most of the ITU-R study groups an important contribution that presented general guidelines to be taken into account when the studies for agenda items were being conducted (Iran (Islamic Republic of), 2020). These principles included, quite significantly, the use of relevant sharing and compatibility studies that have already been conducted. These previously conducted studies, examining a number of frequency bands, have been subject to extensive scrutiny and while there may be some changes due to the use of bands by other services, it is likely the bulk of the conclusions reached remain valid.

The proposed principles also suggested investigating whether or not secondary services should be included in the studies, as well as whether sharing and compatibility studies in one ITU region should take into account its impact on another region. Another suggestion made by Iran was to agree on the sharing and compatibility criteria, assumptions, simulation processes and mitigation techniques before conducting any studies. This was the practice before WRC-19. As it did not occur for WRC-19, several studies were conducted based on different assumptions that resulted in different conclusions being reached.

The Iran contribution was discussed at a meeting of the chairmen and vice-chairmen of ITU-R SGs and WPs. This resulted in a document – 'Principles to be considered and taken into account in studies relating to WRC-23 agenda items' – being drafted. The ITU-R BR Director highlighted that these principles were provided as information and guidance for all the ITU-R groups, with the aim of harmonising as much as possible the work of each group. They do not, however, supersede the main ITU-R resolutions covering the working methods of WPs and SGs (ITU-R Director, 2020).

This contribution, from one of the most influential countries in the ITU-R, deserved attention given that addressed several drawbacks encountered by the international community during the last study period and, especially, before WRC-19. However, these principles are just guidelines – they do not achieve mandatory status until they are accepted by the ITU-R assembly. In addition, the contribution revealed several deficiencies within ITU-R decision-making procedures that will need to be carefully addressed in order to provide more certainty to WRCs decision and facilitate consensus.

Having said that, we have highlighted the need to revise procedures that have been used for more than a century. In particular, voting may be used in future WRC as there are large differences between the requirements of ITU-R countries. Furthermore, it seems that some countries such as the USA have a relatively faster pace of technology development that cannot wait for a WRC decision. This is illustrated by the case of 5G and Wi-Fi6E. There is also a need to incorporate new players in the wireless telecommunications sector (e.g., Microsoft, SpaceX) into the decision-making process. Meanwhile, some countries act as protectors of the RR, resisting any major proposal that impacts on the traditional and existing radiocommunication services within the RR.

3.6 Interference: Certainty versus Innovation

Interference is at the heart of international spectrum management – the RR were established more than century ago on the premise of not causing interference to stations operating according to these regulations. The ITU's historic approach to mitigating harmful interference has been to divide spectrum according to the type of service and to implement the global harmonisation of allocations. Harmful interference is defined in the RR as "interference which endangers the functioning of a radionavigation service or of other safety services or seriously degrades, obstructs, or repeatedly interrupts a radiocommunication service operating in accordance with the RR" (ITU-R, 2020a).

Such an approach has not changed since the RR of 1927 when it was explicitly mentioned that any country should not cause interference to another countries' services (Ard-Paru, 2013). Furthermore, the main radiocommunication services (mobile, fixed, broadcasting) have been in the RR since 1927 with only a few changes. However, it is necessary to ask whether such a traditional paradigm results in a specific path for wireless technologies where there is a development in several areas (e.g., modulation, physical layer and antenna array) but within a context that embraces restrictive interference protection and separation between radiocommunication services.

The reason for such question is that although there have been several developments in wireless technologies in the last century (Tranter et al., 2007), no innovation has emerged that alters the traditional paradigm of interference internationally that focuses exclusively on the protection of existing services regardless of their spectrum utilisation and receiver performance. In particular, an innovation creating a disruptive wireless technology that is highly immune to interference and can achieve a self-spectrum management is unlikely to occur given that industry is directing its R&D expenditure in a way that affirms the existing handling interference paradigm.

Meanwhile, several innovative ideas have been proposed in the literature that address national spectrum management that should be discussed at some point by ITU-R study groups. For instance, there is a suggestion to develop a dynamic model where the parameters of the licence (e.g., bandwidth, power) could be changed by the regulator over time (Saint & Brown, 2019). Within the RR, flexibility could be provided by differentiating between actual spectrum use and service allocation. In such a case, the ITU-R would play the role of an international regulator that specifies some parameters depending on the allocated services that could be changed within limits to enable the flexible use of spectrum at borders. With regard to the management of interference, it has been suggested that alternative measures instead of traditional exclusion zones are applied that are dynamic, multi-tiered and adjust to the interference power (Bustamante et al., 2020).

Another observation is that although the concept of sharing between different services, in terms of primary and secondary to increase the utilisation of the spectrum, has been around for almost a century, the percentage of spectrum sharing primary and secondary allocations is quite small. In 2012, the figure was almost 10% (Ard-Paru, 2013). This indicates that the concept of spectrum co-sharing is, in practice, difficult to implement due to the resistance of exiting incumbent primary services. In other words, when new or existing services providers want to gain access to a specific part of the spectrum, whether on a primary or secondary basis, they have to prove that they will co-exist with current services. These conditions imposed are based on the assumptions that these current services are in full operation and require full protection. The conditions may be unrealistic and similar to windfall profits in spectrum auctions as they stimulate current services to continue instead of moving to another spectrum band. This has a negative impact on new innovative services that need access to the spectrum to provide services that may compete with existing services (e.g., proprietary MTC vs. LTE-M).

Furthermore, on the national level, there has been a concern that long-term exclusive licences restrict innovation (Lehr, 2005; Standeford, 2018). There is a need to address whether rapid technological development has created a similar situation internationally where service allocations are fixed in large bandwidths for a period of decades. It is, quite simply, difficult to change such allocation or introduce an additional allocation. So how should existing services be encouraged to move if they do not need or fully utilise their allocated spectrum? Individual countries can motivate existing operators (e.g., broadcasters) to release their spectrum in return for financial compensation by holding incentive auctions (e.g., FCC auctions in the UHF). But how can we achieve that internationally?

A first step is to be able to assess the utilisation of different existing services, by conducting a survey of all ITU-R countries or by utilising monitoring stations as it may be difficult to assess passive services. For passive services, there should be an agreed way to assess their need for spectrum. For instance, prior to WRC-19, there were several accusations from bodies like NASA that commercial 5G deployments in the 24 GHz will interfere with weather observation satellites that gather data for forecasts and hurricane prediction (Hollister, 2019). However, it was clarified by the mobile industry that the main sensor to be impacted, causing the loss of 70% of weather forecasting data, namely the Conical scanning Microwave Imager/Sounder, was cancelled in 2006 (Gillen, 2019).

Regarding the application of incentive auctions internationally, within the ITU momentary principles are not applied: instead, functional and technical studies are considered by WRCs. However, an auction could be applied to international spectrum management with some caveats and in a different way than on the national level (e.g., SMRA). For instance, WRC could apply a holistic approach to all radiocommunication services where countries submit their requirements in a specific band. In such approach, footnotes are not conditioned by the agreement of neighbouring countries but by whether the use is justified and does not affect neighbouring countries. WRCs would act in such case as the facilitator of exchanging spectrum between the different services and countries, while the auction addresses the demand of the different services taking into account the utilisation of existing services and sharing conditions. But how can we introduce financial consideration of the opportunity cost of introducing new radiocommunication services? One way is to have independent economic assessment studies undertaken by the World Bank, another UN organization, to be included in the CPM text prior to WRCs so that countries have insights into the economic implications of their decisions.

To this end, is it possible to hold an incentive auction for broadcasting services in the 470–698 MHz band in Region 1? This is an example of several services competing to access the spectrum. Firstly, it should be noted that there are two treaty agreements covering this band within Region 1. The first is the RR, which accommodate the different service allocations in the band, whereas the second is the RRC agreement that covers terrestrial broadcasting services. While the RR could be altered through a WRC agenda item, the RRC agreement could only be modified by another RRC. However, another RRC with the same procedures may result in the same output. It is, therefore, suggested to hold a regional conference (e.g., RRC-27) but not only for broadcasting as mobile services also need to be included. One group of countries can have a mobile allocation, and another group of neighbouring countries can have broadcasting services where there are some stringent conditions on the borders of these spectrum bands.

The difference between previous RRCs is that countries will submit their actual requirements, and the approval of the requirements of other countries would not be conditional on the sole approval of neighbouring countries but instead by having reasonable conditions. During such a conference, there will be several iterations similar to auction rounds, where each country re-submits their requirements. When the conference ends, mobile allocation in a specific geographic area starting from 570 MHz and broadcasting allocation starting from 470 MHz in other geographic areas would have been agreed. This would provide pressure on both mobile and broadcasting services to utilise new technologies and provide a second chance to countries to select the appropriate services to meet their specific needs.

Emerging from this is the need to extensively review a priori planning. An interesting question to ask is what would have been the result of a planning conference in the UHF band if it would be conducted now instead of in 2006? Would countries have the same requirements that they anticipated two decades ago, when most of world uses their mobile services for simple functions (such as texting and downloading small data). We doubt that this would be the case as RRC-06 was based on estimated maximum requirements for each country rather than realistic ones. Countries sought to acquire as many TV channels as possible that could be deployed in the future, but given the lag experienced by many countries regarding their digital switchovers the number of channels sought would overestimate future demand.

Another question that can be asked is why have there been multiple generations of mobile but not broadcasting technologies? In other words, would a priori planning stimulate innovation? Technological developments have occurred in the broadcasting sector, but the pace has been considerably slower than it has for mobile technologies. Furthermore, while there are means to revise historical service allocations, there is no clear or timely way to revise plans like GE-06, which would require another RRC to revise it. In other words, more than 100 countries would need to agree that a revision of the plan was required in order for another regional conference to be held. Reaching such a level of international agreement is not an easy task, and one that could take decades.

It is, perhaps, time to hold an extraordinary WRC to review the main principles covering the whole table of allocation and definitions of different services. This has happened before, in 1947 when the WARC reviewed the whole table of allocation to meet post World War II requirements and also in 1959 when the conference reviewed the whole of the RR (Ard-Paru, 2013). However, holding such a conference would arguably place a strain on the resources available to ITU-R, making it unlikely that such a conference would be held in the foreseeable future. Instead, a more limited conference could be held that would review (a number of) concepts: for example, WRC decision-making processes, the involvement of wider array of actors/stakeholders, traditional service categories and, arguably most important of all, the definition of harmful interference. Without such a review, there is a clear risk that in future the ITU-R will become irrelevant to the development of wireless technologies and satisfying end-user requirements. The RR have been able to accommodate advances in wireless technologies, as most of them are within the traditional paradigm of spectrum management, but a disruptive wireless technology that is (highly) immune to interference and can achieve a self-spectrum management has not yet emerged. Until it does, there is no need to alter the current international institutional arrangements for spectrum management, but the window of opportunity that this creates needs to be grasped.

3.7 Conclusion

This chapter has focused on international spectrum policy. The international regime governing the use and allocation of spectrum is long established, initiated at the start of the 20th century in response to failed communications and a maritime disaster. The evolution of this regime has been shaped by technological progress and the guiding principle of avoiding interference between competing services. While the international regime has sought to provide certainty in the allocation and use of spectrum, this has become increasingly difficult as the number of services (applications) has grown and the membership of the ITU expanded. Not only has the expanded membership of the ITU resulted in the emergence of regional associations, but it has also contributed to the differences that exist within and between the organisation's three regions.

Through drawing on a series of current issues, we have sought to explore the degree to which the international spectrum regime can address uncertainty and become more flexible while still providing the certainty and protection from interference that is needed. At the heart of each of these issues are, in essence, a series of trade-offs that accommodate the competing needs of regional associations and leading countries within the international spectrum management regime. If each of these issues were to be taken in isolation, it could be argued that there is relatively little need to change the international regime. It is, however, the collective nature of the issues that we have discussed in this chapter that, we believe, highlight the need to change the international spectrum management regime.

One way to address this need for change would be to hold an extraordinary WRC to undertake a review of all spectrum allocations principles and service definitions. While this may be attractive, especially given the increasing demands being placed on spectrum and the competing interests of some services and countries, but it would be a difficult, perhaps insurmountable, task to arrange an extraordinary WRC. Alternatively, incremental changes to the allocation of spectrum could be made. The international regime would evolve, with countries participating in a framework with which they are familiar, but this may be too slow given the pressing demands for spectrum. A third option, of course, lies somewhere between these two alternatives. It would involve accelerating the pace at which the ITU-R works, speeding up decision-making processes to reflect the quickening pace of technological change and new spectrum demands, while also being more pragmatic. Although there are resource implications, on the ITU as well as its member countries, of such an approach, it does offer a practical way forward.

References

3GPP. (2019). *3GPP Status on 6 GHz*. WInnForum Workshop on 6 GHz. Washington, DC.

Al-Rashedi, N. (2014). *Existing SRD Related ITU-R Deliverables*. ITU Workshop on Short Range Devices and Ultra Wide Band, Geneva.

Analysys Mason. (2019). *Discussion on the 6 GHz Opportunity for IMT*. A. M. AS.

Anker, P., & Lemstra, W. (2011). Governance of Radio Spectrum: License Exempt Devices. In: W. Lemstra, V. Hayes, & J. Groenewegen (Eds.), *The Innovation Journey of Wi-Fi: The Road to Global Success*. Cambridge University Press, Cambridge, UK.

Ard-Paru, N. (2013). *Implementing Spectrum Commons: Implications for Thailand*. Chalmers University of Technology, Sweden.

Arefi, R., & Cordeiro, C. (2020). *Spectrum Needs of Emerging License-Exempt Technologies*. Retrieved from https://builders.intel.com.

ASMG. (2015). *Arab States Common Proposals. Common Proposals for the Work of the Conference. Agenda Item 1.1*. World Radiocommunication Conference (WRC-15), Geneva.

ATU. (2015). *African Common Proposals for the Work of the Conference. Agenda Item 1.1*. World Radiocommunication Conference (WRC-15), Geneva.

Benkler, Y. (1998). Overcoming Agoraphobia: Building the Commons of the Digitally Networked Environment. *Harvard Journal of Law and Technology, 11*(2), 1–113.

Bhatia, B. (2020). *Wi-Fi in 6 GHz*. National Workshop on New Wi-Fi Spectrum in 6 GHz. Retrieved from https://itu-apt.org/.

Bustamante, P., Das, D., Rose, J. S., Gomez, M., Weiss, M. B., Park, J.-M., & Znati, T. (2020). *Toward Automated Enforcement of Radio Interference*, TPRC48: The 48th Research Conference on Communication, Information and Internet Policy, TX, USA.

CEPT. (2000). *WRC-2000: European Common Proposals for the Work of the Conference*, European Commission, Brussels.

CEPT. (2015). *European Common Proposals for the Work of the Conference, Agenda Item 1.1*. World Radiocommunication Conference (WRC-15), Geneva.

Cisco. (2020). *Cisco Annual Internet Report (2018–2023) White Paper*. Retrieved from www.cisco.com.

Coleago Consulting. (2020). *The 6GHz Opportunity for IMT.* Retrieved from www.coleago.com/.

Downes, L. (2012). Requiem for Failed UN Telecom Treaty: No One Mourns the WCIT. *Forbes.* Retrieved from www.forbes.com.

DSA. (2021). *How to Realise the Full Potential of 6 GHz Spectrum.* Retrieved from http://dynamicspectrumalliance.org/.

ECC. (2014). *ECC Report 224: Long Term Vision for the UHF Broadcasting Band.* Retrieved from https://docdb.cept.org/.

ECC. (2019). *ECC Report 302: Sharing and Compatibility Studies Related to Wireless Access Systems Including Radio Local Area Networks (WAS/RLAN) in the Frequency Band 5925–6425 MHz.* Retrieved from https://docdb.cept.org/.

ECC. (2020). *ECC/DEC/(20)01 Decision on the Harmonized Use of the Frequency Band 5945–6425 MHz for Wireless Access Systems Including Radio Local Area Networks (WAS/RLAN).* Retrieved from https://docdb.cept.org/.

Egypt, Jordan, Lebanon, & Morocco. (2015). *World Radiocommunications Conference 2015: Proposals for the Work of the Conference. Agenda Item 1.1.* Retrieved from www.itu.int.

El-Moghazi, M. (2017). *A Game of Frequencies at WRC-15: The Future of ITU-R at Stake?* PTC-17, Honolulum, USA.

El-Moghazi, M., & Whalley, J. (2019). *IMT Spectrum Identification: Obstacle for 5G Deployments.* TPRC, Washington, DC.

Emirates News Agency. (2020). *TRA Adds Additional 500 MHz of 6 GHz Band for the Wi-Fi Radio Frequency Spectrum.* Retrieved from www.wam.ae.

Ericsson. (2019). *Agenda Item 10.* 3rd Working Group Meeting for WRC-19, Gaborone.

EU. (2017). *Decision (EU) 2017/899 of the European Parliament and of the Council of 17 May 2017 on the Use of the 470–790 MHz Frequency Band in the Union.* Retrieved from https://eur-lex.europa.eu.

FCC. (2020). *FCC Opens 6 GHz Band to Wi-Fi and Other Unlicensed Uses.* Retrieved from www.fcc.gov.

Finland. (2015). *World Radiocommunications Conference 2015: Proposals for the Work of the Conference. Agenda Item 1.1.* Retrieved from www.itu.int.

Frieden, R. (2019a). The Evolving 5G Case Study in Spectrum Management and Industrial Policy. *Telecommunications Policy, 43*(6), 549–562.

Frieden, R. (2019b). *WRC-19 and 5G Spectrum Planning.* TPRC47, Washington, DC.

Gillen, B. (2019). *How a Fake Weather Sensor Could Take Out 5G.* Retrieved from ctia.org.

Gross, D. A., & Lucarelli, E. (2011). *The 2012 World Conference on International Telecommunications: Another Brewing Storm Over Potential UN Regulation of the Internet.* Regulatory Communications 2012, USA.

GSMA Intelligence. (2020). *Intelligence Brief: Have 6 GHz Decisions been Hasty?* Retrieved from www.mobileworldlive.com/.

Hazlett, T., & Leo, E. (2010). *The Case for Liberal Spectrum Licenses: A Technical and Economic Perspective.* George Mason Law & Economics, Research Paper No. 10–19, Virginia, USA.

Hemingway, D. (2020). *WRC-23 Agenda Item 1.5 A Broadcaster's Perspective.* UK SPF Event on WRC-23 Agenda Item 1.5.

Hollister, S. (2019). *5G Could Mean Less Time to Flee a Deadly Hurricane, Heads of NASA and NOAA Warn.* Retrieved from www.theverge.com.

Industry Canada. (2011). *Review of the International Telecommunication Regulations (ITRs) 2012 World Conference on International Telecommunications (WCIT-2012).* Retrieved from www.ic.gc.ca.

Internet Society. (2012). *What are The ITRs?* Retrieved from www.internet society.org.

Iran (Islamic Republic of). (2020). *Contribution to Working Parties 4A, 4B and 4C on Relevant Agenda Items of WRC-23.* Working Party 4A. Retrieved from www.itu.int.

ITU. (2006). *Article 1: Definitions.* Final Acts of the Regional Radiocommunication Conference for Planning of the Digital Terrestrial Broadcasting Service in Parts of Regions 1 and 3, in the Frequency Bands 174–230 MHz and 470–862 MHz (RRC-06).

ITU. (2012). *WCIT-12 Overview.* Retrieved from www.itu.int.

ITU-R. (2001). ITU-R Recommendation SM.1265–1: National Alternative Allocation Methods. In *SM Series. Spectrum Management,* ITU, Switzerland.

ITU-R. (2003). *Resolution 229: Use of the Bands 5 150–5250 MHz, 5250–5350 MHz and 5470–5725 MHz by the Mobile Service for the Implementation of Wireless Access Systems Including Radio Local Area Networks.* Provisional Final Acts – World Radiocommunication Conference (WRC-03).

ITU-R. (2006). *Recommendation M.1767: Protection of Land Mobile Systems from Terrestrial Digital Video and Audio Broadcasting Systems in the VHF and UHF Shared Bands Allocated on a Primary Basis.* ITU-R Recommendations M-Series.

ITU-R. (2007a). *Article 5: Frequency Allocations,* ITU, Switzerland.

ITU-R. (2007b). *Resolution 805. Agenda for the 2011 World Radiocommunication Conference.* Provisional Final Acts – World Radiocommunication Conference (WRC-07).

ITU-R. (2007c). *Resolution 953: Protection of Radiocommunication Services from Emissions by Short-Range Radio Devices.* WRC-07 Final Acts.

ITU-R. (2007d). *Resolution 956. Regulatory Measures and Their Relevance to Enable the Introduction of Software-Defined Radio and Cognitive Radio Systems.* Provisional Final Acts – World Radiocommunication Conference (WRC-07).

ITU-R. (2012a). Article 5: Frequency Allocations. In *Radio Regulations: 2012 Edition* (Vol. 1), ITU, Switzerland.

ITU-R. (2012b). *ITU-R Resolution 58: Studies on the Implementation and Use of Cognitive Radio Systems.* Retrieved from www.itu.int.

ITU-R. (2012c). *WRC-12 Recommendation 76 Deployment and Use of Cognitive Radio Systems*, ITU, Switzerland.

ITU-R. (2012d). *WRC-12 Resolution 232. Use of the Frequency 694–790 MHz by the Mobile, Except Aeronautical Mobile, Service in Region 1 and Related Studies.* Provisional Final Acts – World Radiocommunication Conference (WRC-12).

ITU-R. (2014a). *Report BT.2302: Spectrum Requirements for Terrestrial Television Broadcasting in the UHF Frequency Band in Region 1 and the Islamic Republic of Iran.* ITU-R Reports BT-Series.

ITU-R. (2014b). *Report ITU-R BT.2339–0: Co-Channel Sharing and Compatibility Studies Between Digital Terrestrial Television Broadcasting and International Mobile Telecommunication in the Frequency Band 694–790 MHz in the GE06 Planning Area.* ITU-R Reports BT-Series.

ITU-R. (2014c). *Report ITU-R M.2330–0: Cognitive Radio Systems in the Land Mobile Service.* ITU-R Reports M-Series.

ITU-R. (2015a). Article 5: Frequency Allocations. In *Radio Regulations* (Vol. 1), ITU, Switzerland.

ITU-R. (2015b). *ITU-R Recommendation M.2090 : Specific Unwanted Emission Limit of IMT Mobile Stations Operating in the Frequency Band 694–790 MHz to Facilitate Protection of Existing Services in Region 1 in the Frequency Band 470–694 MHz.* ITU-R Recommendations M-Series.

ITU-R. (2015c). *Resolution 160: (WRC-15) Facilitating Access to Broadband Applications Delivered by High-Altitude Platform Stations.* WRC-15, Geneva.

ITU-R. (2015d). *Summary Record of the Twelveth Plenary Meeting.* World Radiocommunication Conference (WRC-15). Retrieved from www.itu.int.

ITU-R. (2015e). *WRC-15 Resolution COM 6/2: Agenda for the 2023 World Radiocommunication Conference.* Provisional Final Acts – World Radiocommunication Conference (WRC-15).

ITU-R. (2016). *Invitation for Submission of Proposals for Candidate Radio Interface Technologies for The Terrestrial Component of The Radio Interface(s) for IMT-2020 and Invitation to Participate in Their Subsequent Evaluation.* Retrieved from www.itu.int.

ITU-R. (2017a). *Report BT.2337: Sharing and Compatibility Studies Between Digital Terrestrial Television Broadcasting and Terrestrial Mobile Broadband Applications, Including IMT, in the Frequency Band 470–694/698 MHz.* ITU-R Reports BT-Series.

ITU-R. (2017b). *Report M.2415: Spectrum Needs for Public Protection and Disaster Relief.* ITU-R Report M-Series.

ITU-R. (2018a). *ITU-R Recommendation M.2003–2: Multiple Gigabit Wireless Systems in Frequencies around 60 GHz.* ITU-R Recommendations M-Series.

ITU-R. (2018b). *Report F.2438: Spectrum Needs of High Altitude Platform Stations (HAPS) Broadband Links Operating in the Fixed Service.* ITU-R Reports F Series.

ITU-R. (2018c). *Report ITU-R F.2438–0: Spectrum Needs of High-Altitude Platform Stations Broadband Links Operating in the Fixed Service.* ITU-R Reports F-Series.

ITU-R. (2018d). *Report M.2440: The Use of the Terrestrial Component of International Mobile Telecommunications (IMT) for Narrowband and Broadband Machine-Type Communications.* ITU-R Reports M-Series.

ITU-R. (2019a). *CPM Report on Technical, Operational and Regulatory/Procedural Matters to Be Considered by the 2023 World Radiocommunication Conference.* CPM-23-1, Sharm El-Shiekh.

ITU-R. (2019b). *Recommendation M.1036–6: Frequency Arrangements for Implementation of the Terrestrial Component of International Mobile Telecommunications (IMT) in the Bands Identified for IMT in the Radio Regulations.* ITU-R Recommendations: M-Series.

ITU-R. (2019c). *Report M.2480–0: National Approaches of Some Countries on the Implementation of Terrestrial IMT Systems in Bands Identified for IMT.* ITU-R Reports M-Series.

ITU-R. (2019d). *Resolution 1–8: Working Methods for the Radiocommunication Assembly, the Radiocommunication Study Groups, the Radiocommunication Advisory Group and other groups of the Radiocommunication Sector.* ITU-R Resolutions.

ITU-R. (2019e). *Resolution 241: (WRC-19) Use of the Frequency Band 66–71 GHz for International Mobile Telecommunications and Coexistence with Other Applications of the Mobile Service.* WRC-19, Sharm El-Sheikh.

ITU-R. (2019f). *Resolution 245: WRC-19) Studies on Frequency-Related Matters for the Terrestrial Component of International Mobile Telecommunications Identification in the Frequency Bands 3 300–3 400 MHz, 3 600–3 800 MHz, 6 425–7 025 MHz, 7 025–7 125 MHz and 10.0–10.5 GHz.* WRC-19, Sharm El Shiekh.

ITU-R. (2019g). *Resolution 646: (Rev. WRC-19) Public protection and Disaster Relief.* WRC-19, Sharm El-Shiekh.

ITU-R. (2020a). Article 1: Terms and Definitions. In *Radio Regulations,* ITU, Switzerland.

ITU-R. (2020b). Article 4: Assignment and Use of Frequencies. In *Radio Regulations,* ITU, Switzerland.

ITU-R. (2020c). Article 18: Licenses. In *Radio Regulations,* ITU, Switzerland.

ITU-R Director. (2020). *Outcome of the Sixteenth Meeting of the Chairmen and Vice – Chairmen of the Radiocommunication Study Groups, Working Parties, and other Subordinates Groups.* WP 5D.

Jones, D. (2017). Wi-Fi Spectrum Needs Study Released. *IPass Blog.* Retrieved from www.ipass.com.

Kholod, A. (2020). *CEPT Begins the Build-up to the 2023 World Radiocommunication Conference.* Retrieved from http://apps.cept.org/.

Lamy, P. (2014). *Report to the European Commission: Results of the Work of the High Level Group on the Future Use of the UHF Band (470–790 MHz).* Retrieved from ec.europa.eu.

Lehr, W. (2005). *The Role of Unlicensed in Spectrum Reform*. Massachusetts Institute of Technology, MA, USA.

Marti, M. R. (2019). WRC-19 Identifies 4.8 GHz for IMT in Surprise Move. *PolicyTracker*. Retrieved from www.policytracker.com.

Ofcom. (2020). *Improving Spectrum Access for Wi-Fi: Spectrum Use in the 5 GHz and 6 GHz Bands*. Retrieved from www.ofcom.org.uk.

Paoletta, P. (2019). *World Radio Conference Outcomes*. Retrieved from www.europeaninstitute.org/.

RCC. (2015). *Common Proposals by the RCC Administrations on WRC-12 Agenda Item 1.1*. World Radiocommunication Conference (WRC-15), Geneva.

Royabalat, A. (2020). *Importance of 6 GHz to Wi-Fi Ecosystem*. Radio Spectrum for IMT-2020 and beyond: Fostering Commercial and Innovative Use.

Saint, M., & Brown, T. X. (2019). A Dynamic Policy License for Flexible Spectrum Management. *Telecommunications Policy, 43*(1).

Sbeglia, C. (2021). Saudi Arabia Designates Entire 6 GHz Band for Unlicensed Use, Paving Way for Wi-Fi 6E. *rcrwireless*. Retrieved from www.rcrwireless.com.

SFCG. (2020). *SFCG Objectives for WRC-23*. Retrieved from www.ioag.org/.

Standeford, D. (2018). Longer Spectrum Licences don't Necessarily Drive Operator Investment. *PolicyTracker*. Retrieved from www.policytracker.com.

Sung, L. (2003). Observations from WRC-03. *International Journal of Communications Law and Policy* (8).

Tranter, W. H., Taylor, D. P., Ziemer, R. E., Maxemchuk, N. F., & Mark, J. W. (2007). *The Best of the Best: Fifty Years of Communications and Networking Research*. W.-I. Press, USA.

USA. (2015). *World Radiocommunications Conference 2015: Proposals for the Work of the Conference. Agenda Item 1..1*. Retrieved from www.itu.int.

U.S. Congress Office of Technology Assessment. (1993). *The 1992 World Administrative Radio Conference: Technology and Policy Implications*, US Government Printing Office, Washington, DC, USA.

Wallsten, S. (2020). *Ambassador Grace Koh on WRC-19 and Spectrum for 5G (Two Think Minimum)*. Retrieved from https://techpolicyinstitute.org.

Webb, W. (2018). *There's a Standards Problem in IoT, Here's How to Solve It*. Retrieved from https://internetofbusiness.com.

Weyl, G. (2020). *How Market Design Economists Helped Engineer a Mass Privatization of Public Resources*. Retrieved from https://promarket.org/.

Wi-Fi Alliance. (2019). *Agenda Item 10*. 3rd Working Group Meeting for WRC-19, Gaborone.

World Radiocommunication Report. (2015). Proponents Pressed For Flexibility on a Country-by-Country Basis in Lower UHF Contingent on Agreement with Neighbors, Existing GE06 Provisions. *World Radiocommunication Report, 6*(43).

Youell, T. (2015). WRC-15 Agrees on "No Change" Until at Least 2023 for 470–694 MHz in Region 1. *PolicyTracker*. Retrieved from www.policytracker.com.

4

INNOVATION AND SPECTRUM MANAGEMENT

An Oxymoron?

WILLIAM WEBB

Contents

Innovation is much like motherhood and apple pie – very few would deny it is a good thing and most would support more of it. Digital connectivity is, arguably, one of the most fertile areas for innovation, with dramatic changes in the way we use broadband and mobile communications over the last two decades. With radio spectrum being one of the key inputs into the provision of mobile communications, it is unsurprising many regulators have explicit duties to further innovation, and all would wish to be seen as supporting it. However, this chapter will argue that they rarely proactively help, and mostly hinder, albeit not intentionally. We start by considering the problem, before looking at some case studies and then turning to possible solutions.

The Problem

A good place to start is with a definition. One commonly used definition of innovation is:

> Innovation is the successful exploitation of new ideas.

This suggests two parts to innovation – the first that it is something new and the second that it turns into a success, which normally means

DOI: 10.1201/9781003156765-5

85

it delivers profit. Mobile telephony was innovative, as was video calling, social media and much more. The Internet of Things could be argued to be innovative in the home and office where adoption is widespread but perhaps not yet in the wider world where it is still to be successfully exploited.

Innovation appears to happen "naturally" in a capitalist society. There are strong incentives to innovate as this can lead to very substantial monetary rewards. There is also a human desire to create and to make a difference. Many start-ups are launched – fewer succeed. Broadly, there is no shortage of good ideas but plenty of barriers to their successful deployment.

The regulator is one such barrier. Because demand exceeds supply then access to spectrum needs to be limited, so almost by definition regulators will be blocking some innovation. Regulators may also impose some non-spectrum barriers such as net neutrality, but these are not considered further here. Regulators should be more concerned about removing any barriers to innovation than about stimulating it.

Of course, some innovations can occur within the current spectrum framework. Mobile operators can launch new services over existing networks (but rarely do[1]). New technologies and services can be introduced into unlicensed spectrum where the rules of access allow. Many regulators provide test-and-trial licences that allow for temporary deployment of new technologies. But there are innovative ideas that do not fit within the current framework. This may be because, for example:

- Spectrum rules prevent certain types of deployment, such as maximum power limits in unlicensed spectrum making outdoor IoT deployments unviable due to limited cell range and hence high expense.
- Licensing approaches effectively preventing certain types of deployment, for example if spectrum is auctioned nationally this effectively prevents local deployment, such as of 5G networks within a factory by the owner.
- The new idea would cause interference to current users, for example low-orbit satellites interfering with fixed links, and the regulator decides to block or severely curtail the new idea.

In cases such as this, the regulator must make a decision. Simplistically, will they favour the status quo, or will they change the rules to enable the new usage? We argue that most have a strong bias towards the status quo and hence they effectively block many innovations. There are many reasons for this bias including:

- Regulators are inherently risk-averse and so favour the status quo and incumbents. Most innovation comes from new entrants.
- Regulators are prone to "regulatory capture".
- Regulatory processes such as consultation – especially when there is an international dimension – are very slow which makes it difficult for those who need funding from venture capitalists and similar, which requires relatively rapid routes to exit.
- Regulators have rarely been in start-ups and so fail to understand the issue.
- An evidence-based approach is inherently anti-innovation as there is no evidence for new ideas and typically little demand or representation to the regulator.

Many of these reasons are subjective and likely would be refuted by regulators, so it is worth spending time discussing the more subjective of them.

Most important of all is the view that regulators are inherently, indeed institutionally, risk-averse. If this is so then clearly they will be inclined more towards the status quo, and so counter to innovation (unless that innovation comes from the status quo). Partly this observation is based on personal experience over the last 30 years. But it also makes logical sense. Regulators are Governmental entities. They are not profit-driven. Governmental bodies are subject to scrutiny from audit authorities and the public, often aided by freedom-of-information legislation and increasingly subject to legal challenge on major decisions. It is well known that public sector careers are furthered by being a "safe pair of hands" and can be rapidly destroyed by making clearly visible errors. There is rarely any reward in the public sector for making a bold decision that proves correct, but substantial down-side from making a poor decision, or even a good decision that results in adverse publicity. Risk-aversion can also act as a veto. It can only require one risk-averse individual to halt a process, especially if they are senior or have specialist knowledge (e.g., a lawyer). It is inconceivable (and probably inappropriate) that a regulator would have a motto such as "move fast and break things". Indeed, it is hard to conceive of a regulator of even having a motto.

This view that regulators are institutionally risk-averse is not a criticism but an observation. Risk-aversion is entirely appropriate in many cases – clearly no risk should be taken with spectrum used for services such as air-traffic control, public safety or even mobile broadband. If a choice had to be made between risk-aversion and innovation then few would doubt that the right answer would be risk-aversion. Of course, as we will explore later, it need not be an either-or choice.

Another subjecting assessment is that regulators are subject to regulatory capture. This is a well-researched phenomena summarised[2] as:

> Regulatory capture is an economic theory that says regulatory agencies may come to be dominated by the industries or interests they are charged with regulating. The result is that an agency, charged with acting in the public interest, instead acts in ways that benefit the industry it is supposed to be regulating.

There is nothing to suggest that spectrum regulators are immune, and indeed, given the dominance of a few large organisations such as the mobile operators, and their dedicated lobby and liaison capabilities, regulatory capture would seem likely. The occasional movement of regulator staff into regulatory roles in industry adds to the likelihood of this sort of behaviour.

The final element worth discussion is the problems caused by a desire to be evidence-based. On the face of it, making evidence-based decisions sounds like an excellent idea and much better than making decisions based on gut-feel. However, for decisions that are forward-looking in that they enable new services to be introduced over the coming decade, there is generally little evidence available, especially for services that do not yet exist. Requiring evidence will invariably favour the status quo, where evidence is available and forward-extrapolation (e.g., into future data demands) more tractable. In this case, a better approach is to turn to independent experts who have shown a good track record of forecasting the future.

If all of this is true, then there is clearly a problem with regulators and innovation. But if this is so why has this not been more widely acknowledged? Broadly it is because it is impossible to say whether a regulator is stimulating innovation, to assess past performance or to quantitatively rank because:

- Blocked innovations do not become visible and so demonstrate what could have happened had the regulator been more open to innovation.
- Innovative ideas spread globally making country-by-country comparison impossible – a country that is more open to innovation effectively provides the evidence that allows other countries to then permit the service.
- Other factors such as the availability of VC funding and national culture will play a big role in how well a country performs and hence the regulatory effect is just part of a larger picture.

In the next section, we discuss some case studies that indicate regulators blocking innovation.

Case Studies

Case studies are inherently hard to come by, as they are studies in what did not happen. Here we look at some situations where innovation might have happened but appears to have been blocked by regulators. There is no way of knowing whether new services would have succeeded had the regulatory outcome been different, but at least they would have had the opportunity to try to succeed.

TV white space. A number of companies interested in deploying innovative new services noted that there was substantial free spectrum in the TV band (roughly 470–790 MHz with national variations) that could be accessed using dynamic database systems which would protect the incumbents. Different services were proposed including national IoT networks that enabled long range, long battery life and ample capacity. Regulators started considering allowing TVWS access in 2008, but it was only finally allowed in the USA in 2012 and the UK in 2014. Worse, even when allowed it was subject to the

UHF band being reduced in size to provide more spectrum for mobile communications, leading to substantial risk for those aiming to use it. Most national regulators decided not to allow it often citing "insufficient evidence of demand". For the start-ups trying to exploit it, a four to six-year regulatory process proved far too long to sustain funding and most went out of business. Regulators saw this as further reason not to open up the bands. As a result, TV bands remain under-utilised and few countries have national IoT networks. IoT connections have only reached around 20% of the levels that forecast by 2020.

What could regulators have done differently? They could certainly have moved much faster – there was no reason for this process to have taken longer than a year even allowing for studies and consultation. They could have given more certainty of access to the bands even if used for mobile operations, for example by reserving a few channels for unlicensed operation. They could have worked collaboratively on an international basis to deliver greater harmonisation.

Sharing in the 5G C-band. The 3.4–3.8 GHz "C-band" is seen globally as a key band for 5G deployments. But many note that some of the foreseen 5G applications would benefit from self-deployment, for example in factories, offices or on campuses. Innovative spectrum award processes could allow for this, for example, through sharing of spectrum where the operators were not using it and some set-aside for unlicensed use or similar (the German regulator has set aside some of the band for industrial usage). In the UK, the Government department seeking to promote 5G (the Department of Digital, Culture, Media and Sport or DCMS) recommended that the regulator allow shared access to the band. But the regulator declined, preferring a standard auction process which it felt would be simpler and faster. This was despite growing evidence from the USA where shared access to the band under the CBRS initiative was providing popular.

What could regulators have done differently? They could have looked afresh at cellular technology, understanding that the deployment scenarios had changed and hence the licensing process should change too. They could have studied more carefully the work needed to allow shared access and hence the extent of any delay. They could have been less inclined to align with the mobile operators understanding the regulatory capture likely to have occurred.

LightSquared/Ligado. This is a US case study with a nearly 20-year history. While LightSquared only emerged in 2010, its predecessor dates from 2001. The companies have been seeking to use spectrum adjacent to the GPS band for a 4G/5G mobile network but have needed FCC approval. This has been long-delayed due to concern that interference might occur into GPS receivers. Only in 2020 was approval finally provided, by which time LightSquared had been through bankruptcy and re-emerged as Ligado. Even now, this approval is being challenged. While there are interference questions that need study, there is no reason why this could not have been completed in a year. A decade or more is clearly very detrimental to innovation and suggests major process issues. Clearly, the regulator could have moved much more quickly and should learn lessons as to how to cut timescales below a year in future, including changes to processes if needed.

Common threads.

1. *Lack of pace.* Regulators move slowly, innovation requires speed. Regulatory processes should enable all decisions to be implemented within a year, and regulators should be pressured to ensure that this happens.
2. *Fear of the new.* Regulators imagine anything new will take substantial resource and much time. But with focus and appropriate use of external resource, most of the new ideas can be implemented quickly.
3. *Preference to incumbents.* Most of the regulators have a tendency to assume that incumbents should have first preference on existing and new spectrum.

Solutions

We have suggested that innovation is generally desirable but that spectrum regulators will inherently be risk-averse and biased against innovation. This is no bad thing – disruption to safety-critical or widely used spectrum services would be highly problematic. This implies that no one-off fix will work as the regulator will then just lapse back into business as usual.

We suggest that an "innovation czar" could help – an individual within the regulator with responsibility for seeking out new ideas, liaising with those leading the innovation and representing them at senior management level in the regulator. This would work best if the innovation czar were on a time-limited contract so that they did not tend towards the cultural norms of Governmental employees. Their role would be easier if a substantial fraction of the senior management group had prior experience in start-ups or similar innovative entities, and this should be one of the "diversity" criteria when appointing senior managers.

The innovation czar should provide an annual report on how the regulator has promoted innovation and the cases where innovative ideas had been blocked or delayed. This should be public and available for consultation and comment.

Often other Government departments have more of a remit to encourage innovation – for example to promote national companies. A close working relationship with such departments and a bias towards assisting them could encourage innovation. Regulators tend to resist this as it reduces their independence but used with care it will likely bring more benefits than risks.

A process review may be appropriate. For example, some decisions may not need consultation, or there may be a role for expedited consultations where delay can hinder innovation.

Perhaps, above all, if regulators recognise that despite their warm words, they are institutionally anti-innovation, they would be more alive to the risks this runs and willing to put in place mitigation measures.

Notes

1. While mobile operators would argue that they are innovative, it is hard to think of a successful new idea that they have enabled, beyond their original introduction of mobile telephony (when, of course, they were a new entrant). While they have trialled video telephony, walled garden internet, "widgets" as a form of apps, m-health, m-wallets, location-based services and so much more, they are very rarely successful in delivering them. However, as a (non-innovative) provider of a critical service they are unsurpassed.
2. See, for example: www.investopedia.com/terms/r/regulatory-capture.asp.

5

HOW SHOULD WE USE DEMAND FORECASTING IN SPECTRUM POLICY?

RICHARD WOMERSLEY

Contents

The one thing that is absolutely certain about predictions is that the probability of them being correct is infinitesimally small. If it was possible to forecast anything with any accuracy, there would be no lotteries (as it would be possible to predict the results), there would be no betting on the outcome of sporting events and indeed there would be no gambling industry at all.

Predictions are generally based on one of three premises:

- What might be deemed a bottom–up approach in which the various parameters that affect an outcome are carefully modelled to yield an entirely theoretical outcome.
- An empirical approach in which measurements of known parameters are made, equations are designed which yield as close as possible results to the measurements, and then these are used to extrapolate other values.

DOI: 10.1201/9781003156765-6

- A hybrid of the two= in which theoretical values are tweaked to try and replicate results which are found through measurements.

Examples of these approaches can be seen in the way in which radio coverage is predicted, and the pros and cons of each can be clearly demonstrated.

Propagation Models

The free space path loss (Friis) equation is a bottom–up model which determines the extent to which a radio signal will be attenuated (reduced) as it travels through a vacuum, uninterrupted by any obstacles whatsoever. It is based on pure physics and is, as it happens, 100% accurate in situations where radio waves travel through vacuums. Even in the depths of outer space, however, there are occasional hydrogen atoms hanging around, whose presence cannot be predicted and thus there are no actual circumstances where the equation will produce a completely correct result. Here on planet Earth, there are far too many obstacles that would impede a radio signal such as the atmosphere, the ground and the fauna and flora which inhabit it, and as such the Friis model is almost completely guaranteed not to produce the correct answer, except in the case of the occasional fluke.

The Okumura–Hata path loss model is purely empirical and is based on trying to fit equations around a set of signal strength measurements taken in Tokyo such that the results of the equations match the measurements as closely as possible. Its inaccuracies, therefore, come in various different forms:

- As the equations aim to emulate the results of the measurements, they will inevitably produce results which are close to, but not the same as, the actual values. Thus, even in Tokyo where the original data was measured, the model will not give results which fully match reality.
- As the input data were measured in Tokyo, the model will only give results which are close to reality in cities whose parameters (e.g. geography, topography or population density) are very similar to those of the Japanese capital. In practice,

the only place that will closely match these parameters is Tokyo and as stated earlier, the model is not even truly accurate there.

The Longley–Rice path loss model is a hybrid. A bottom–up model has been varied based on measured statistical variances to improve the accuracy of the underlying physics. The model even allows users to state the required level of statistical accuracy, such that the result can be determined for, for example, 50%, 75% or 90% of the time. To achieve coverage for 75% of the time requires a higher average signal strength than it would to achieve coverage for 50% of the time such that any fluctuations in signal (or error in calculation) can be overcome by the additional received power. Likewise to achieve coverage for 90% of the time requires more average received signal strength than would be needed to achieve reception 75% of the time. In essence, the model adds in an increasing safety margin to overcome its inaccuracies and of the vagaries of radio propagation itself. If the exact signal strength which would be present at any point, and at any time, could be accurately forecast, then the precise amount of signal needed to enable successful reception could be employed and no additional margins would be necessary.

Which of these three approaches is the most appropriate (or indeed, the most accurate) depends upon a number of factors. Firstly, what amount of data is available. The Friis model requires only knowledge of the frequency of operation and the distance between transmitter and receiver. Okumura–Hata additionally requires knowledge of the heights of the transmitter and receiver above ground level. And the Longley–Rice model also requires terrain and other topographical information.

Secondly, the level of accuracy desired. In outer space, no one would scream at the use of the Friis model, which is probably sufficient for most of the purposes, and for rough estimates of mobile propagation on the Earth, something akin to the Okumura–Hata model may well be sufficient. It also depends upon what the calculations are being used for. Coverage prediction, interference prediction, compatibility analysis and cross-border coordination may all demand different levels of accuracy, and an appropriate model needs to be selected. The same logic, and same basic premises, apply to the calculation of spectrum demand, and it is incumbent upon the modeller, and those relying on its outputs, to properly understand available input data, and the purpose to which its output is going to be put.

Calculating Spectrum Demand

The International Telecommunication Union's (ITU) model for determining spectrum demand[1] for International Mobile Telecommunications (IMT) is undoubtedly one of the most sophisticated models in the field. It is an example of a bottom–up approach. It fuses together data on the demand for IMT services across a range of different geographic areas (i.e. cities and rural areas) with assumptions about usage density, spectrum efficiency and cell sizes to reach a conclusion on the amount of spectrum required. However, despite its complexity, it has yielded results that have proven to be less than accurate. In 2014, the ITU published a report[2] predicting spectrum demand for IMT in 2020. This report forecasts that a total of between 1340 and 1960 MHz of spectrum would be required by 2020. By 2019 (prior to WRC-19), the total amount of spectrum identified for IMT services by the ITU was around 1300 MHz, however, the average of spectrum amount across the world licensed by regulators to mobile operators in their country was nearer 500 MHz, despite data growth exceeding that forecast by the ITU.

Clearly, the model had a number of flaws. One of these was an estimate of the data consumption per user, which forecast that by 2020 in urban areas, this would exceed 200 Gbytes per user per month, whereas, for comparison, in 2019 the average data consumption per mobile user did not exceed 10 Gbytes per user per month.[3] This is a factor of 20 difference, whereas the outputs of the model are less than a factor of 5 out of kilter with reality, which leads to the obvious realisation that other inputs to the model must also have been inaccurate.

At the other end of the scale, in 2010, the Federal Communications Commission (FCC) forecast that an additional 275 MHz of spectrum would be required for US mobile networks to meet expected traffic demand by 2014.[4] This was an empirical model, based on taking the actual values of various parameters as they were in 2009, and extrapolating how they would change over the ensuing 5 years. In reality, over that period, US wireless operators were able to accommodate all of the traffic growth projected, without deploying even the spectrum that was already allocated for wireless services in the USA by 2010, despite the fact that both traffic growth and the number of cell sites deployed were largely in line with the FCC's projections. The difference was primarily that spectrum efficiency outpaced that which the FCC expected. Other models such as those developed for the Australian Communications and Media Authority and for the GSMA which use similar but perhaps more sophisticated empirical processes have shown similar levels of inaccuracy.

Whilst no one would expect any predictions to be 100% accurate, and hopefully those reviewing and cogitating on the results of such predictions would understand the inherent difficulties of predicting the future, it is perhaps the level of inaccuracies that is staggering. A forecast which was wrong by a factor of 2, stating that demand was 100 MHz when in reality it should be 50 MHz could easily lead to decisions being taken on spectrum allocation which yielded an inefficient outcome. When the values produced (and the inputs used) are out by factors many times higher than this, it is easy to begin to understand why predictions should not just be treated with caution, but with, perhaps, a healthy degree of scepticism.

Timescales

Spectrum policy requires an understanding of the relative demand for frequencies from all of the services which need access to the valuable, limited resource. Decisions over who should access which frequencies, and when, are taken internationally at the World Radiocommunications Conference (WRC) which usually takes place once every 4 years. For a matter to be discussed at the WRC, it has to be put on the agenda at the preceding WRC meaning that there is a minimum 4 years lag between raising an issue concerning spectrum allocation and the first date at which an internationally recognised decision might be taken. If you take into account the time required for standards bodies to agree on a technological approach for a service, it may take 8 to 10 years from an initial proposal for a new radio service to an international decision being made on providing it with spectrum. Even at a national or regional level, the process of allocating radio frequencies can be a time-consuming and laborious process.

It is, therefore, necessary for all industries to produce future-looking forecasts of their likely demand for spectrum in order to take decisions on which service should be allocated which frequencies and, just as importantly, when it will be needed. It is not just the amount which is needed that should be determined, but it is also essential to have a feel for when spectrum will be needed as this provides the timeframe in which any existing users can be re-farmed into alternative bands in time for the new service to gain access. And here we have somewhat of a vicissitude. On the one hand, predictions of spectrum demand are important to make spectrum available, but these forecasts are often generated many years in advance of the demand becoming reality. For a new service, such as cellular was in the 1980s, the forecast of demand for radio spectrum had to be made before the service had even been launched.

Demand Forecasting Inputs

It is worth noting that any calculation for spectrum demand will be based on a series of underlying assumptions concerning a number of elements, such as:

- the number of users of a service;
- the extent to which those users will exercise the service (e.g. how long they will spend on phone calls or how much data they will use);
- the required quality of service (e.g. how often, when a user goes to make a call, or stream a video, will the network be busy);
- the necessary penetration of coverage (is it outdoor only, or must indoors be covered too, and to what degree of certainty);
- how usage is distributed (i.e. is it primarily during one or two busy hours or at a few hotspot locations, or spread evenly across time and geography);

- the spectrum efficiency of the technology in use, and how this varies by location (e.g. the difference between the centre and the edge of a cell); and
- the density of the wireless infrastructure, as more infrastructure requires less spectrum.

As was earlier discussed, simple inaccuracies in any of these factors can lead to significant errors in a forecast's results. Any forecast of the demand for radio spectrum cannot, therefore, be isolated from the need to forecast the demand for the service itself. The history of radio spectrum allocation is littered with examples where forecasts of demand have been woefully inaccurate, if not downright wrong. Here are a couple of examples.

In the 1980s, pagers were all the rage. Based on growing demand for paging services, in 1990 European Council Directive 90/554/EEC designated frequencies for the introduction of a pan-European land-based paging service. This directive forced every European Union Member State to set aside radio frequencies which would then be used, Europe-wide, for a new paging service. Unfortunately, those lobbying for the paging spectrum had totally failed to foresee that SMS would rapidly overtake pagers as the *de facto* means of sending short messages. Eight years later, a 1998 study by CEPT showed that the paging service was not developing and that manufacturers had ceased producing the necessary equipment. The spectrum designated for the paging service, therefore, sat largely fallow for nearly 15 years until in 2005, European Parliament and Council Directive 2005/0147 repealed the 1990 directive freeing up the spectrum for use for other services.

Whilst it could be argued that those developing the paging service could not have reasonably predicted the simplicity and convenience of the use of SMS, this example serves to illustrate how forecasts based on past growth turned into predictions of future growth can be grossly misguided.

In 1998, when Glick GSM was successful in bidding for the second mobile licence in Egypt, the incumbent operator Mobinil had around 85,000 subscribers. Click's business case was predicated on the assumption that within the first 5 years the Egyptian mobile market may reach 2 million subscribers, whilst some industry observers suggested that this may top 4 million. Their forecasts of likely uptake were partially based on the expected cost of the service compared with the country's levels of disposable income. By the end of 2003, the number of mobile subscribers had reached almost 6 million,[5] far ahead of even the wildest forecasts.

The roar away success of the mobile service in Egypt compared with the initial forecasts was not due to any particular miscalculation by the operators but by the fact that the official figures on incomes, based on the Government's receipt of taxes, significantly underestimated the 'grey economy' and the amount of disposable cash that was sitting in peoples' pockets. In addition, the fact that having a mobile device was highly fashionable helped divert cash from other expenditure towards the mobile networks.

The longer the period between any known facts against which a forecast can be calibrated and the prediction itself, the greater the likelihood of inaccuracies. As an example, in the 1970s, AT&T, which was lobbying the US government to provide radio spectrum for a cellular service, had estimated that by the turn of the 21st century there might be upwards of 2 million subscribers in the USA. In reality, 30 years later, there were over 90 million. Forecasts of service demand should therefore be treated with as much of a modicum of scepticism as the spectrum demand forecasts themselves, especially given the historical inaccuracies which have blighted future predictions.

It is not just the models used to predict spectrum demand which should be scrutinised but also the range of inputs. It was George Fuechsel, an early IBM programmer who is said to have coined the phrase 'garbage in, garbage out' and the notion that trusting any kind of result which is based on erroneous inputs should not be trusted is as true today as it was in the 1950s.

In telecommunications, it is common to consider the demand placed upon a network during the 'busy hour'. This is generally defined to be the hour of the day during which the highest demands are placed on the network. The busy hour can be different across geographic locations. At a commuter railway station in the centre of a busy city, for example, the busiest hours will be the morning and evening rush hours when the station is packed with commuters. In a dormitory commuter town, the busiest hours might be during the school run in the morning and later in the evening when the commuters return from the cities.

The principle of the busy hour is that if the network is able to sustain all of the traffic generated during this time, then it will loaf along easily for the rest of the day. The demand for data instead of voice calls has changed the time distribution of traffic and has generally resulted in lower, more sustained peaks, however, the principle still holds that if you can support the traffic during the period of highest demand, then the rest of the time the network will not be stretched.

The question, however, is what qualifies as a reasonable figure for the busiest time? Is it a standard working day, or should it include the demands of subscribers around periods such as the New Year when everyone wishes to have a video call with their loved ones? For a rural location, is it the daily demands of residents, or the peak caused by a packed commuter train passing through the coverage of the cell?

In many of the spectrum demand models, the needs of those working in dense urban office environments have been taken to be the benchmark. If these users' needs can be satiated, then in less dense areas users should be fine. But which city should be used? The population density of the Earth is around 15 people per square kilometre, but given the large amount of almost unoccupied space (such as deserts), this would clearly not reflect actual land usage. Conversely, Manila is the most densely populated city in the world with an average of around 43,000 residents per square kilometre. Even in a city such as this, there will be hotspots of increased density. Areas with tall housing or office blocks concentrate people into far smaller areas. In the UK, for example, there are 872 square kilometres of the country where the residential population density exceeds the 43,000 average for Manila.[6]

Too Much Spectrum Is Also a Problem

The previously discussed model used by the ITU to calculate demand for IMT, for example, uses the density of office workers in an undefined, busy downtown Asian city (such as Tokyo) for its demand predictions. The population density value used is 220,000 people per square kilometre. Whether or not this is a valid figure for Tokyo or Tianjin, the question has to be asked as to whether using an input such as this will give results that should be used globally to determine the demand for spectrum. Whilst this might represent the busiest, busy hour on the planet, the mobile operators of densely populated Asian countries presumably account for this by massively increasing the density of their base stations in these heavily populated areas. And certainly, the fact that network densification takes place in these areas can also be (and indeed with respect to the ITU model is) another input to the spectrum demand model. Thus, spectrum demand in the ITU model is based on dense population and dense network deployment. Whatever the outcome of such modelling, should this value also be used to determine spectrum demand in downtown Ouagadougou, or even central Paris? On the one hand, it could be argued that a worst case situation based on the busiest of the busy is, as discussed earlier, a peak figure and that the rest of the world's mobile networks could theoretically loaf along with the amount of spectrum needed to satiate demand in central Tokyo. However, the role of the policymaker is to assess relative demand and determine who should have access to how much spectrum. Overinflated demand may lead to decisions which deny access to other services, or reduce the amount they have.

Too little spectrum tends to lead either to poorer quality services or to more expensive ones; however, it is also worth bearing in mind that too much spectrum can be just as much of a headache as not enough. Give an operator in a small, sparsely populated country, the same spectrum as those serving central Shenzhen and they may wonder what to do with it. They may not have the investment funds available to roll-out services in each and every frequency band, and their demand may never reach the point where they need so much spectrum. And in the process, someone else who would otherwise be using that spectrum may have lost out increasing the cost, or lowering the performance of their services.

How Can We Assess Models?

Given that the spectrum is a limited resource with a wide range of competing demands, it is in the nature of those who are seeking access to spectrum to overestimate their needs. Ask the mobile operators in any country for their forecast of their future market penetration and if you add them all together it will always, impossibly, exceed 100%. Spectrum demand forecasting is no different: Ask users to forecast their demand and add it all together and once again the total will exceed 100% of that which is available.

How can this issue be resolved? Initially, a careful examination of the assumptions made in preparing the forecasts is useful, however, not every organisation is willing to provide details of the inputs and assumptions they have made. Sometimes (rarely) this is because the input data may contain confidential information such as that provided by mobile operators. More often, unwillingness to share details falls into two categories:

- The model has been prepared by a consultant whose model is covered by their intellectual property rights, and as such they do not wish to make it public. This is reasonable as much time and effort can be put into developing a model which can then be re-used for multiple clients or purposes. You would not expect Microsoft to freely release the source code for Excel, for example.
- The values used might not stand up to independent scrutiny. This might particularly be the case where wide-ranging assumptions have been made which, although reasonable, could be subject to significant variance, thus resulting in very different outcomes which the model's user might not wish to be revealed. As previously discussed, a factor of 2 difference can be very significant and any input whose value could reasonably be challenged could change the results by this much or more.

Another element to consider is the motivation of the author of the forecast. In most cases, they will be seeking to gain access to more spectrum and, therefore, it might be expected that they will over-estimate the actual requirement. This is largely understood by decision makers. Of course, the same party might wish to downplay the demands of another sector, in favour of their own, in which case it might be expected that they are likely to have underestimated that requirement. When a report is published by a particular industry body, it is generally obvious as to what their likely positioning is, however, when renowned international organisations publish documents, it might be expected that their outputs are more balanced.

Even organisations such as the ITU, whose reports might be expected to be impartial can be subject to subconscious bias. Many of the working parties who are tasked with investigating a particular industry are staffed by those with a vested interest in that industry. ITU Working Party 5D which is tasked with examining IMT issues is thusly staffed largely by mobile manufacturers and other parties with an interest in the promotion and development of the mobile industry. Similarly, Working Party 6A whose focus is terrestrial broadcasting is primarily attended by, and contributed to by, those from the terrestrial broadcasting industry. Being an internationally renowned organisation does not therefore guarantee even-handedness, instead it hopefully ensures that those working on a particular topic are experts in the field.

Notwithstanding all of the this, forecasts of spectrum demand form an essential input into the decision-making processes that take place at both a national and an international level with regards to which services should be allocated how much spectrum. This begs the question of how regulators should respond to the various models and results presented to them in order to take sensible decisions as to the use of spectrum.

The Loudest Voices

One suggestion is that the old premise of 'the empty vessel makes the loudest noise' applies. The original meaning of this is that those who have the least knowledge or the least to say are often those who make the most noise. However, in this case, it might be expected that those whose glass is nearly empty, insofar as they are running low on available spectrum, would shout the loudest. This is often the case for those who are seeking to preserve the spectrum they have. As their service becomes squeezed into ever less spectrum, a breaking point comes at which they realise that if they do not take action, any further reductions in the amount of spectrum they have will begin to materially damage their service. Some industries are better at this kind of advocacy than others. The terrestrial television broadcasters, the programme making (PMSE) community and the satellite industry have all been very vociferous in their protection of the remaining spectrum they have. Conversely, the defence community has been relatively quiet, despite having had many of the bands they historically used heavily, re-allocated for other uses. That being said, there is a quiet acceptance that certain bands, being military in nature, are off the cards when it comes to shaking up their usage. The same might be said for aeronautical and certain other uses, where it is recognised that any additional interference may pose a threat to safety-of-life and as such, those pushing for new spectrum take a more hands-off approach and so in these cases screaming does not necessarily equate to pain.

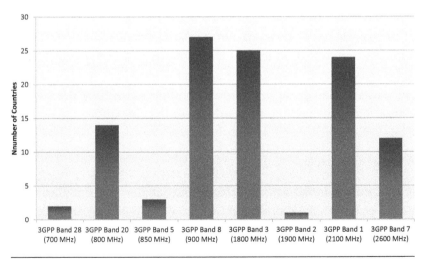

Figure 5.1 Licensing of mobile bands across 28 African countries.

Source: LS telcom.

In a different vein, however, the mobile industry has been equally vociferous in their push for ever more spectrum, despite not having actually put much of the spectrum they have previously gained at various WRC's to use. The 2.6 GHz band, for example, originally known as the '3G expansion band', was identified for IMT usage at the WRC in 2000 and yet out of 28 African countries for which data were available at the time of writing, only 14 have, 21 years later, licensed the band (known as 3GPP Band 7) to mobile operators. Indeed only 21 have awarded spectrum in the 2.1 GHz band (known as 3GPP Band 1) despite this being identified for mobile services in 1992.

The mobile industry's argument is that, based on the ever growing demand for mobile data and the sloth of the international frequency allocation processes, it is imperative that they are provided access to more spectrum now, in anticipation of the demand they forecast in the future. It may well be the case that in Japan, South Korea and other highly developed countries, such spectrum demand will come to pass; however, global decisions continue to be driven by these extreme cases which do not reflect the worldwide pattern of usage. It is worth remembering that each time a new frequency band is identified for a new service, it is often the case that the incumbent user has to be re-farmed to an alternative band which introduces new headaches for regulators who often have to undertake a lengthy process with multiple public consultations, to clear the band for the new use.

Whilst, therefore, those who make the loudest noise may well have strong reason to seek additional spectrum, or protect that which they have, sometimes it is simply the amount of resources that a particular industry focuses on spectrum defence (or attack) that determines whose voices are heard most clearly above the chatter.

A Geographic Balance

It is often the broadcasting, defence and satellite communities who have lost out to the mobile industry in arguments over who should have how much spectrum. Mobile services have very clearly demonstrated their capability to help with the economic development of countries; however, this ought not to be at the detriment of a loss of connectivity in other areas, and it need not be. What is good for one country may be bad for another and under the ITU structure of the international spectrum decision-making process, the world is divided into three longitudinally defined regions and countries with very disperse and diverse requirements are artificially lumped together. China and Japan have the same spectrum allocations as Tuvalu and Bhutan; Italy and Sweden have the same spectrum allocations as Cabo Verde and the Seychelles; and yet clearly their spectrum demands are going to be very different.

This issue should not be underestimated. The satellite community press hard to protect their use of C-band (3.4–4.2 GHz) from encroachment by the mobile industry, primarily for 5G services. However, in many countries, the use of C-band for satellite services is almost an afterthought, a hang-over from bygone days. With the exception of a few hub stations operated by national telecommunications companies for connection to their overseas territories, or by satellite operators themselves, it would be very unusual to find C-band being used heavily in many countries. In countries with heavy rainfall, however, the use of C-band satellite is absolutely critical, providing connectivity even during the heaviest storms. However, the countries where heavy rainfall exists, and those where C-band satellite use may be relatively rare, are within the same ITU region and at an international level end up with the same frequency allocations. Such a position can lead to unwise decisions being taken by regulators. Their desire to find spectrum for 5G services and the growth of the ecosystem in C-band (known as 3GPP bands n77 and n78) means that regulators seek to re-allocate the band from satellite to mobile services. What often goes unnoticed is that the same frequencies are in widespread use for, for example, satellite television broadcasting. Introduce mobile services into the band and all of those watching television will suddenly find themselves with no picture, or at best a badly broken one. The fact that these downlinks are not registered does not mean that they do not need protection and there have been instances where, having begun to launch a service in the band, regulators have had to backtrack and revoke licences as widespread interference occurred.

No Simple Solution

Is there a way that balanced decisions can be taken which balance the needs of different users and the needs of different countries? In the simplest terms, probably not. Whilst country-specific footnotes can be incorporated into the ITU's frequency allocation regulations, there is a sense in which countries in a given region do not want to be left behind. When companies such as Google tout the notion of using balloons in the stratosphere to provide internet connectivity to unserved areas, everyone wants to be on-board and spectrum for the High Altitude Platform Stations (HAPS) is allocated across whole regions, whether such a service will ever be launched there or not. It would be foolish for a regulator to turn down such a service, would it not, even if by doing so, access to the spectrum for an existing service is denied. It could even be argued that this is a case where less developed countries may force a decision on more developed countries (where the HAPS service finds less utility), rather than the usual vice versa situation.

And so we are brought almost full circle. Whether it's pagers or stratospheric balloons, finding a path that treads a happy medium between the demands of various services whilst protecting the necessary interests of incumbents is always going to be difficult. Whilst spectrum demand forecasts are a useful tool in pointing a possible trajectory, they should be treated as one might treat an old route map that's been sat in the glove compartment of an aging car for many years: whilst it may accurately show the locations of the main cities, it is unlikely that the roads it shows are still the best way to get from A to B.

Notes

1. ITU-R Recommendation M.1768 "Methodology for calculation of spectrum requirements for the terrestrial component of International Mobile Telecommunications".
2. ITU-R Recommendation M.2290 "Future spectrum requirements estimated for terrestrial IMT".
3. Source: www.statista.com/statistics/489169/canada-united-states-average-data-usage-user-per-month/.
4. www.gsma.com/spectrum/wp-content/uploads/2012/03/fccmobilebbforecastoct2010.pdf.
5. www.statista.com/statistics/497203/number-of-mobile-cellular-subscriptions-in-egypt/.
6. Source: CDRC Maps (maps.cdrc.ac.uk).

6

THE REALPOLITIK OF SPECTRUM IN THE GLOBAL ECONOMY

SIMON FORGE

Contents

6.1 The Realpolitik of Spectrum – a Financial, Industrial and Political View of the Asset

With its historical links to politics and special interests, spectrum has progressively grown into more of a commercial asset than a public resource for the benefit of all.[1] Its net economic importance is likely to grow further over the next century as fixed line telecommunications become progressively replaced by radio communications to an extent not yet seen, although this is beginning with the proposed

DOI: 10.1201/9781003556765-7

rollouts of fixed wireless access (FWA) for the last link in the local loop.[2] Successful low-cost medium range mobile broadband will drive that. Such migration may be seeded by the mobile operators and also perhaps by the latest Low Earth Orbit (LEO) communications services and microsatellites industry, but they in turn may be ultimately replaced by the online platform players (especially as they already dominate global subsea cable expansion and soon will lead the next generation of earth linked space communications via LEOs). But the interpretation of the future of spectrum should take in a higher-level view than that of the supply side operators and increasingly, their supply chains – for the semiconductor components, equipment, software and handset devices. Future views should include the global competition between regional and national economies and their industrial strategies which may include complex long-term planning of new generations of technologies, as in Korea[3] and Taiwan to some extent (TSMC – or the Taiwan Semiconductor Manufacturing Company – started as a government-funded enterprise).

This influence is due to the power of spectrum to define markets and market access for the key operator players via a *frequency defined* system of signal discrimination, in contrast to licence exempt (LE) swathes. The fundamental commercial impact of that is spectrum's ability to provide a mechanism for exclusion of others and so to set the competitive landscape. Allocations of frequency bands to technologies and services, followed by assignment of specific frequency slots to unique operators defines today's radio-based economy. For market entry, a unique licence is necessary. So spectrum makes markets.

Furthermore, frequency division offers the licence holder an effective form of market foreclosure, justified on the basis of pollution of its own signals by others' interference. Spectrum, as always, is the arbiter of who can enter and compete in markets that contain some of the most profitable sectors on the planet. To a most important extent for open innovation, that exclusive licenced model has been countered by the licence exempt bands with various sharing mechanisms (e.g. Listen Before Talk or LBT). This spectrum has afforded key innovations, leading to the Wi-Fi ecosystem, Bluetooth linking and the instrument, scientific and medical bands, which have seeded complete sectors for ISM devices as well as logistics and transactions RFID, at frequencies from VHF, UHF on into the millimetric bands.

Moreover, there is a question facing the economy, whether acknowledged or not, that exclusive licences for management of the spectrum market mechanism introduce the *costs of infrastructure competition*. This has been successful in the past and continues to be so.

Whether it can be maintained in the future is an interesting question. That leads to a second evident question – where to go next when all current bands are allocated. It might lead to increasingly higher spectrum licence values and a search in apparently empty bands – which are probably empty for good reason.

That is most true when there is a contradiction between the spectrum ranges being chosen and the subsequent costs of the network due to the physics of propagation of the chosen frequency bands. A current prime case is the rollout of 5G NR at 'mid-band' frequencies (possibly 1–4 GHz) which may have to be 'piggybacked' in 5G NSA mode on the existing infrastructure for LTE-A via a software upgrade, even in the 'lower' C-band regions. It is for this reason that the 'mmWave' millimetric frequencies (20–30 GHz) at auction form a separate market to any UHF or C-Band lots, yielding much smaller revenues, or are left unsold. It is for this reason, that other methods of spectrum management are important, especially sharing among users[4] which in dynamic form (DSS) also forms a key technique for extending the range limits of 5G technology due to the primary spectrum bands used.

Spectrum use is increasingly driven by the needs of the mobile industry rather than other users, such as broadcast, whose spectrum demands have effectively shrunk with a new form of the 'Negroponte switch' – perhaps the 'Pelton switch'[5] – as their content moves to the web and then over the web via mobile and also Wi-Fi tethering. In contrast, recently the calls of business for radio-based networks have expanded, despite largely being met by the industrial networks in the LE bands[6] for production, logistics and even 'smart' supermarket shelving.

In terms of spectrum assets expansion, mobile cellular has been the lead telecommunications sector over the last three decades, as well as in revenues and utility for the populace, both consumer and business.

Its gains have moved up from hundreds of MHz in the 1990s to MHz and GHz today in regional and national allocations. That is driven by trends in communications by migration to spectrum-based media (online social networking, entertainment, news and now includes business videoconferencing, cloud-based office processes over mobile broadband).

The fixed line fibre optic network gets progressively higher in capacity but mobile connectivity still increases exponentially in user demand. The arrival of a state of civilisation termed *Telecomia*, first expected in the mid-1990s, has triggered a new economic migration – away from urbania and suburbia to a distribution of society not based on proximity to the workplace for the first time in human history. Consequently, spectrum affordances could potentially stimulate a massive reversal in real estate attractiveness and the price of land.

6.1.1 State of the Global Telecommunications Industry Reflects Growing Use of the Spectrum

The mobile services industry and its supply side have survived as fairly stable structurally since its founding in the 1980s, despite frequent mismanagement. While the number of mobile network operators (MNOs) globally has gone up and down a little – the equipment supply chain has changed to become dominated by Asian manufacturers, although the same equipment/component suppliers may persist, they have been joined by lower-cost mass-production to meet demand:

- This stability has survived technology transitions from analogue to digital, over six generations of technology.
- Use of radio communications in products has expanded with reconfigurable radio systems (RRS) in consumer products, industrial machines and processes, and increasingly, in road and rail vehicles using short range devices (SRDs).
- Vertical integration between services and infrastructures has changed little, in the mobile industry despite the influx of MVNOs to 'borrow' the spectrum of MNOs. Shared wholesale use of spectrum on a fixed infrastructure has not, so far, flourished.

- The mobile Industry has optimistically embarked on a search for value in two new fields – firstly, IoT in the factory and the home; secondly, migration to media content for subscription video on demand (SVOD) for the consumer, as the revenues in consumer and business services are stagnant. The former is difficult and may fail, the latter is still being persistently pursued although in the UK, BT with its EE mobile arm, is abandoning some premium sports.
- The design of mobile radio systems continues to be set by the supply side – not governments, standards organisations or the operators – the supply side controls the main regulatory goals and the technical design agenda, for instance, via the 3GPP.

The supply side of equipment, semiconductor components and software increasingly sets the directions because it has the technical leadership over the operators. But more importantly, it also has access to the credit finance for leveraging its network hardware and software when offered to the MNOs. The new highly successful range of supply side market entrants over the last few decades depend on access to low-cost finance (most often from state-owned credit banks that underwrite national exports) to offer the MNOs customers long-term credit anywhere globally. Such finance from suppliers is essential for rollout of new mobile technologies. Long-term financial leverage through export lending to the telecommunications services industry is a key activity for national high technology industrial policy – especially for major Asian suppliers.

Consequently, MNOs are increasingly dependent on equipment finance from suppliers who offer to postpone repayments until the network is generating cashflow, especially needed if high spectrum fees at auction must be paid for first, for market entry. That shapes a powerful symbiotic relationship for controlling the market. But as soon as the network suppliers have sold the current generation of mobile technology and reaped the profits, a new replacement generation is necessary to churn the market and produce the next replacement revenue stream. Sound familiar?

In the global mobile market, various industrial strategies on the support for the supply side are shown by different countries. Nation states recognise that participation in the global mobile technology market is essential for high growth economies and offers a route to enter that category. The global market for spectrum-based products drives the highest value corporations globally. For that reason, suppliers such as Apple and Samsung have such outsize market pricing, with Apple approaching 3TN\$ (30 April 2021) in smartphone devices, while producers such as Taiwan and Korea have highly focused strategies, for semiconductor components especially. Each type of government tends to have a specific form of symbiotic relationship with its supply industry and the MNOs in the services sector.[7]

6.2 The 5G Experience – Its Spectrum Allocation Impacts Set the Mobile Scene for Future Decades

A key example currently of the relationship outlined earlier between the radio spectrum as the gateway to mobile market entry and the supply industry is the 5G 'race'. The 5G technical concepts have been in development really since around 2012[8] triggered by conditions in the global telecommunications equipment industry. The drive to churn the market comes from stagnant conditions in the telecoms equipment sector as MNOs are still digesting their latest purchase of the last mobile network generation – LTE-A and LTE-A-Pro. The suppliers have sold them LTE-A networks, harvested the proceeds and spent it. Supplier share prices are stagnant or down, especially for the equipment suppliers. So the '5G race' is on to churn the market.

The underlying origins of 5G technology may lie more in the need to churn the market when much of the spectrum is already taken in the UHF region (300 MHz–3 GHz). The response is a move up, to higher spectrum bands, with clearing out of any existing occupants such as satellite services who may not have the same leverage with government as the mobile industry. After all, mobile is *the* mass market technology of the century. Consequently, 5G is a supply side market 'push' rather than a demand side 'pull'. It is fuelled by a combination of industrial policy among some major supplier nations, shaped by the supply side. Note that 5G is NOT a replacement for the previous generation, LTE. 5G attempts a major extension of functionality and performance – but is it in the right direction? It requires a sophisticated and complex, network infrastructure – for instance, a high speed, much wider bandwidth national fibre optic long-distance infrastructure, as part of the core network. Thus, 5G is costly, with capital expenditure (CAPEX) and operational expenditure (OPEX) are perhaps up to three times that of previous mobile generations.[9]

In the debate over 5G, the focus is on the performance promised by the new spectrum bands – data speeds of 10Gbps (and theoretically much more) and circuit delay (latency) of milliseconds and less for response and transmission. It seems the promised nirvana – of radio-based fibre optics speeds for the local loop – is in sight, be it mobile or FWA. But whether the ordinary consumer will notice the difference is a key unanswered question for the industry.

More technically, the prime spectrum choices for 5G spectrum have been in the 3–30 GHz band – centimetric bands, with milli-metric suggestions up to 70 GHz. However, higher frequency means lower range for a given power level. That implies the cell density of Base Transceiver Station (BTS) in the cellular architecture must go up as the radius of propagation goes down. Impacts of the physics of the spectrum range means CAPEX investment goes up non-linearly – and considerably with frequency – with a square law unless the BTS prices reduce quickly with size as they become smaller. That is why there is an emphasis for 5G on 'small cells' and mass-produced lower cost SAWAPs (small area wireless access points, in the EU nomen-clature of the EECC).[10]

Furthermore, mobile networks become increasingly complex with 5G in the presence of the great outdoors and the great indoors. Both pose problems of practical 5G signal reception and transmission in the everyday real environment of attenuators ranging from wet foli-age to ferroconcrete buildings and plaster walling. Real environments impose major challenges at the spectrum ranges originally forecast. Moreover most of the mobile network rollouts begin as a real estate exercise, dependent on locating and contracting sites. Even if many of them are in street furniture, the practical problems of laying out a C-Band (or mmWave) network of small cells are intense compared with the previous generations UHF-band networks and expensive in terms of UPS-type power and backhaul, with edge processing and fronthaul.

In consequence, much higher power for transmission is necessary to overcome the rapidly increasing attenuation at such frequencies, which reduces range from many kilometres in the sub-1 GHz range to a hundred metres or less at the upper mmWave range (20–50 GHz).[11]

One solution is to use solid state multi-element phased arrays (MIMO antennae) as active antenna systems (AAS) which produce narrow beams of sufficient power to extend range. However, the price of using this spectrum band is higher power consumption, perhaps up to three times previous mobile generations[12] with new forms of handset needed, probably having MIMO arrays embedded for the uplink path.

Many spectrum auctions for 5G are taking place globally in 2021, as governments have accepted the supply side thesis lobbied over the last six years as it being the incoming mobile technology for the future and they could be left behind. Perhaps over 50 such auctions from 2018 through 2020 have been completed and more are planned, from Chile to China to the Czech Republic. But is there a compelling reason to move to the current form of 5G technology, particularly with choices of spectrum that restrict its utility?

Increasingly, we see from these 150 or so 5G auctions recently completed or in progress that the engineering of 5G has had to include lower frequencies at 'mid-band' levels, to reduce CAPEX and its associated OPEX, effectively almost abandoning mmWave, despite some being purchased at far lower prices. Instead the move to lower bands for 5G has taken two directions:

- Firstly, the move to what is popularly termed the 'C-Band', taken as (often 3.4–3.8 GHz and up to 4.3 GHz) plus other vacant lower bands be they sub-1 GHz UHF, or UHF above that.
- Secondly, the use of spectrum sharing dynamically (DSS) with lower frequency bands from 600 MHz (e.g. T-Mobile USA) up to 1800 and 2600 MHz in which the UHF band is 'borrowed' for the handset uplink as devices such as smartphones may not have the power to transmit over the distance that the BTS transmitter amplifiers have.[13]

Although 5G may ultimately have much high data volumes, it does not have high security as part of its design – as that was added later.[14]

Questions have been raised over this – a fairly complex subject as its high-speed communications and network architecture offer enhanced possibilities for wide-ranging attacks and data thefts, especially as 5G is the first mobile generation to use virtualised functions widely for network management operations. The locus of management control is in a remote data centre, creating a classic single point of failure (SPOF) which is entirely in software 'in the cloud'. This is perhaps perturbing for what may be a critical infrastructure element, especially when being promoted as the technology foundation for real-time critical IoT networks, such as smart grid management, industrial processes or autonomous vehicles.

Much analysis is now going into 5G security audits, somewhat after the fact, especially in the USA following the widespread intrusions such as SolarWinds and Exchange server compromises. DARPA has multiple projects examining 5G security. The European Court of Auditors is examining the security of 5G networks in Europe.[15] This is a reason for looking to a new generation of radio communications that has designs based far more heavily on inbuilt security, as explored briefly later.

The use of 5G by the MNOs to enter the industrial manufacturing and home controls market is also one of the aims of the ITU sets of use-cases for 5G in its International Mobile Telecommunications 2020 initiative.[16] It envisages high reliability and extreme limits on time constants for network response – in the vision that driverless vehicles will be steered by an external network, 5G. Whether the MNOs have the skills, finance and software expertise to move into a completely different métier of industrial systems integration remains to be seen and returns to the bond markets might be necessary to fund it. Such a venture might be as hopeful as the MNOs moving into media, with examples such as AT&T selling off its US$50 Bn acquisition of DirecTV satellite TV company for US$16.25 Bn five years later.[17]

In summary, there are interesting countervailing forces here – while the suppliers want to churn the equipment and software market, the MNOs may not have reaped all the paybacks from LTE-A rollout and so must borrow to buy 5G spectrum. They may be heavily in debt already and so reluctant to borrow more. Signs of this may be seen perhaps in the recent UK spectrum auction in March 2021 when the reserve price was exceeded but barely and the net result was 6% of the revenues from the 3G UMTS auction in the UK in 2000.[18]

Consequently, a pervasive uncertainty for the creditor banks and the MNOs in some countries is whether 5G could be too expensive to roll out, once high spectrum fees have been paid, because the increases in ARPU from 5G customers might not be enough to finance the total investments. This technology demands major network engineering for the expansion it offers in data capacity and speed. If a lot of finance has already gone into spectrum fees at auction that may starve the 5G rollout of capital without careful maintenance of capital and credit lines. This is the reason for moving to the lower frequencies (and perhaps much lower frequencies, below 1 GHz) and sharing existing LTE holdings in the UHF bands using DSS (1800, 2100, 2600, etc.). The rollout is much cheaper. As ever, spectrum sets the potential breakeven ARPU and income floor.

As a result, 5G may trigger alternative infrastructure ownership models, either sharing both spectrum and physical networks or by separation of services and the network, so players may choose either the networking layer or services.[19] Such revised business models may introduce the concept of 'neutral hosts' – third parties owning and operating networks and shared licensed spectrum as alternatives to the current models of infrastructure competition[20] as Malaysia is proposing with a shared wholesale network.[21] This concept of a specialised 5G network operator/owner, supporting all service providers in a neutral fashion has been entertained by the largest operators globally.[22] Whether the business model depends on such neutral hosts or a lesser form of that with an operator-owned shared network infrastructure is unclear. Certainly, in some countries such as China, the MNOs have already agreed to pool the 5G network construction CAPEX and OPEX to reduce financial risk through sharing a common 5G infrastructure, reducing individual funding significantly.[23]

Essentially, this implies an economic limit exists, set by 5G networks' need for far denser, more complex high technology short range networks – and to some extent set by the price of the spectrum. It may be mitigated by using sharing lower frequency bands with DSS, or repurposing existing lower UHF bands for greater range and lower power consumption as 5G carriers. Note that previous mobile generations have taken 5 to 7 years to take off in that reach a reasonable portion of the population. LTE was first rolled out in 2010[24] and is expected to be the majority technology in many countries in 2030. In contrast, 5G may take much longer to have more than an 'early adopter' following (e.g. among enthusiastic gamers) for 10 to over 15 years. That is, unless a replacement technology for 5G appears over the next decade. Such an outcome is not surprising – the telecommunications industry has shot itself in the technological foot before – femto cells, WAP, ISDN – all despite successes in some countries for some technologies.

However, such factors do not always dissuade governments from pursuing a national 5G initiative. On the contrary, it may encourage them if they consider that 5G presence could improve the attractiveness of the economy to foreign direct investment (FDI). So 5G deployments are under way in Argentina, Brazil, Chile, Thailand, Colombia, Vietnam and Mexico. That may well encourage investments, but a 'showcase' strategy faces problems in 'realpolitik' terms. An additional factor is the role of the local private sector in such countries. The MNOs, particularly when speaking in private, may be reluctant to invest in 5G rollouts, due to existing liabilities for the previous generation (LTE-A-Pro). Moreover, returns from 5G are unknown. The more ambitious applications for real-time industrial and home automation networks under the banner of the Internet of Things (IoT) lack a solid business case, as this is a field already crowded with low-cost existing technologies. The MNOs see that since the late 1980s three decades of unending mobile market growth, in fairly strong demand-led conditions, have delivered over 6 billion customers. Demand for this very different technology is still quite unclear, so without government support, MNOs may be far less confident. Within this context, the equipment supply side still needs to churn the telecoms market for new revenues.

Note that the bandwidth of 5G favours the OTT services providers against the MNOs who may tend to survive only as bit carriers. Most important to the MNOs is loss of customer control through the mobile broadband channel, threatening to transfer increased market power as consumer 'ownership' passes to the online web services platforms. These online platform players also control the 5G mobile handset market with the handset operating system/browser market, giving a leverage that 5G could help them to reinforce.

In summary, from a *realpolitik perspective*, the 5G 'race' has been seeded by conditions in the global telecoms equipment industry emphasising the race is a supply side push rather than a demand side pull and fuelled by industrial policy among the major supplier nations.

6.3 Future Directions – Where Are We Going – Next – and How Do We Get There?

6.3.1 *The Next Decade of Radio Technologies Could Set the Directions for the Century*

Beyond the current mobile generations and the incoming 5G, could be a move to a next generation of radio cellular and mesh technologies, as it is unlikely that spectrum will serve the three theatres of operation – high proximity local (<100 m–1 km) for urban; suburban and longer distance rural (<20 km) with indoor penetration; and thirdly wide area coverage, possibly by LEO satellite connectivity as well as a fibre core network to cope with sparse fixed infrastructure environments. Consequently, existing technologies LTE-A and even UMTS may continue well beyond 2030 with their core network protocols and spectrum allocations. But more practical initiatives over the next decade could set the direction for the rest of the century.

6.3.2 *A '6G' for Mainstream Communications – Processing Power Not Raw Spectrum*

Evolution of UMTS to IP packets as in 5G with massive MIMO antennae, cloudified functions and small cells are unlikely to be sufficient for future communications demands, as are moves to the more esoteric parts of the spectrum in the great quest for unused frequency bands.

The move to high data volume radio for communications universally is most likely to need a far more secure mechanism and platform, firstly in an underlying security architecture with efficient coding. It should have a packet structure shaped by the mobile environment for spectrum efficiency. Thus, focus is likely to be on digital signal processing for bandwidth and away from extending raw spectrum to higher frequencies. Faster data speeds and volumes could come from exploiting processing power in BTS and handsets with reuse and sharing of spectrum using cognitive radio and similar techniques. Three areas might thus be expected – new non-IP packet structure rather than the Evolved Packet Core for greater information density. What is needed is a sane, secure and safe next generation of mobile broadband radio networks, with a real business case that can support global rollout at low cost – what 6G should be.

Consequently, perhaps the most interesting area for 6G would be a new secure software architecture to replace the Worldwide Web and today's Internet for security reasons.[25]

6.3.3 A Next Generation of Networking Must Acknowledge Cybercrime Problem Is Overwhelming Us

With the spread of broadband globally, especially mobile broadband, there has been an explosion of cybercrime problems. Increasingly sophisticated cyberattacks are undermining the functioning of our communications networks, critical infrastructure and services – and now affect our society and economy. For example, the widespread SolarWinds attacks on computer networks in the USA tend to confirm that in-depth cyber-attack exposure[26] is expanding yearly, led by professional intrusion teams.

In consequence, while our future lies in moving to broadband radio-based networks, they need to be redesigned for protected business operations, consumers' security and privacy, safe government networks and secure mobile financial transactions for all. Today, our basic communications fixed infrastructure is weak in its cybersecurity fundamentals, and the radio tails offer an even weaker yet ubiquitous gateway for intrusion. The move to 5G can only open further security issues.

6.3.4 Abbreviations and Acronyms

5G NR	Fifth Generation New Radio	LTE	Long Term Evolution (of UMTS)
AAS	Active Antenna System	LTE-A	Long Term Evolution – Advanced
ARPU	Average Revenue Per User	LTE-A-Pro	LTE-A-Professional
BTS	Base Transceiver Station	MIMO	Multiple Input Multiple Output (active antenna array)
DSS	Dynamic Spectrum Sharing	MNO	Mobile Network Operator
DTT	Digital Terrestrial Television	MVNO	Mobile Virtual Network Operator
EC	European Commission	NFV	Network Function Virtualization
EECC	European Electronic Communications Code		
EU	European Union	OTT	Over the Top
FWA	Fixed Wireless Access	RFID	Radio Frequency Identification
IoT	Internet of Things	SAWAP	Small Area Wireless Access Point
IP	Internet Protocol	SDR	Software Defined Radio
ISDN	Integrated Services Digital Network	UHF	Ultra-High Frequency (300 MHz–3 GHz)
ISM	Instrument, Scientific and Medical	UPS	Uninterruptible Power Supply
LAN	Local Area Network	VHF	Very High Frequency (30–300 MHz)
LBT	Listen Before Talk	WAP	Wireless Access Protocol
LE	Licence Exempt	Wi-Fi	Wireless Fidelity – a radio LAN standard
LoS	Line of Sight		
LEO	Low Earth Orbit (satellite)		

Notes

1. T.W. Hazlett, 2017. *The Political Spectrum*. New Haven, CT, USA: Yale University Press.
2. European Commission, DG CNCT, 2018. *Quality and Performance Indicators for Fixed and Mobile Convergence in Europe*. Smart 0046–2016, IMIT, Belgium.
3. S. Forge and E. Bohlin, 2007. Managed Innovation in Korea in Telecommunications – Moving Towards 4G Mobile at a National Level. *Telematics and Informatics*. doi:10.1016/j.tele.2007.10.002.
4. EC-2 European Commission, 2012. *Perspectives on the Value of Shared Spectrum Access*. DG CNCT. https://ec.europa.eu/digital-agenda/sites/digital-agenda/files/scf_study_shared_spectrum_access_20120210.pdf.
5. J.N. Pelton, 1992. *Future View; Communications and Technology in the 21st Century*. Baylin Pub. Corp.

6. I. Amin et al., 2018. *ZigBee Protocol, IEEE 802.15.4.* Science Direct. www.sciencedirect.com/topics/engineering/zigbee-protocol.

7. ITRE Committee, 3 September 2019. *5G State of Play.* Presentation, European Parliament, Brussels.

8. 5G PPP Architecture Working Group, July 2016. *View on 5G Architecture.* https://5g-ppp.eu/wp-content/uploads/2014/02/5G-PPP-5G-Architecture-WP-July-2016.pdf.

9. G. Yang and H. Shujing, 14 November 2019. Will China Lead the World With 5G Network-Sharing Model? *Caixin Business & Tech.* www.caixinglobal.com/2019-11-14/will-china-lead-the-world-with-5g-network-sharing-model-101483449.html.

10. European Commission, DG CNCT, 2018. *Light Deployment Regime for Small-Area Wireless Access Points (SAWAPs).* Smart 2018–0017, Final Report, Belgium.

11. M. Rumney, 2020. *Bridging the Gap Between 5G Vision and Reality.* Cambridge Wireless, Presentation, At 5G: A Practical Approach. www.cambridgewireless.co.uk/resources/radio-by-moray-rumney-from-keysight/.

12. J. Xie, 10 October 2020. Chinese 5G not Living Up to its Hype. *VOA News.* www.voanews.com/east-asia-pacific/voa-news-china/chinese-5g-not-living-its-hype.

13. 5G Networks, 2020. *5G Dynamic Spectrum Sharing (DSS).* Posted on July 2020 by 5g Networks. www.5g-networks.net/5g-technology/5g-dynamic-spectrum-sharing-dss/.

14. ETSI, July 2020. *5G Security Architecture.* TS 133 501 – V16.4.0 – ETSI, 3GPP. www.etsi.org/deliver/etsi_ts/133500_133599/133501/16.04.00_60/ts_133501v160400p.pdf.

15. ECA, 2020. *European Court of Auditors, Audit of Implementing 5G Networks in the EU and its Member States,* 08 December 2020, Press Release. www.eca.europa.eu/Lists/ECADocuments/INAP20-14/INAP_5G_Security_EN.pdf.

16. ITU-R, November 2017. *Minimum Requirements Related to Technical Performance for IMT-2020 Radio Interface(s)* (PDF). IMT-2020 Standard. ITU. www.itu.int/dms_pub/itu-r/opb/rep/R-REP-M.2410-2017-PDF-E.pdf.

17. LEX, 4 May 2021. Verizon/Apollo Media. *Financial Times.*

18. N. Fildes, 18 March 2021. Spectrum Sale Scrapes Past Reserve. *Financial Times.*

19. M. Marti, 4 February 2019. MNOs Starting to Consider Shared Mobile Networks? *Policy Tracker.*

20. Small Cell Forum, December 2017. *Vision for Densification into the 5G Era: Release Overview.* Document 110.10.01. www.scf.io/en/documents/110__Vision_for_Densification_into_the_5G_Era_Release_Overview.php

21. Malaysian Wireless, 25 February 2021. *Malaysia Govt to Build & Manage its Own RM15bil 5G Network with TM and Huawei.* www.malaysianwireless.com/2021/02/malaysia-govt-build-own-5g-network-telekom-malaysia-huawei.

22. F. Grijpink, T. Härlin, H. Lung, and A. Ménard, 2019. *Cutting Through the 5G Hype: Survey Shows Telcos' Nuanced Views.* www.mckinsey.com/industries/telecommunications/our-insights/cutting-through-the-5g-hype-survey-shows-telcos-nuanced-views.

23. K. Kawase, 23 August 2019. Chinese Mobile Carriers Eye Sharing 5G Networks to Curb Costs. *Nikkei Asian Review.*

24. Top Mobile, 2010. *Nationwide 4G/LTE Rollout for TeliaSonera in Norway and Sweden.* Posted January 13, 2010. Top Mobile Accessories. https://topmobileaccessories.wordpress.com/2010/01/13/nationwide-4glte-rollout-for-teliasonera-in-norway-and-sweden/.

25. *Policy Tracker*, 19 March 2021. *As We May Communicate.*

26. H. Murphy, 18 December 2020. Cyber-Attack Poses Ongoing Grave Risk to US Business and State Networks. *Financial Times.*

7

A NEW STAKEHOLDER PARADIGM TO LINK 6G WITH SUSTAINABLE DEVELOPMENT GOALS AND SPECTRUM MANAGEMENT

MARJA MATINMIKKO-BLUE

Contents

7.1 Introduction

This chapter will address the role of sustainability and spectrum access in the context of the next generation of mobile communication networks after 5G – namely 6G. The deployment of 6G systems is targeted for 2030, which is also the target year for the achievement of the United Nations' Sustainable Development Goals (UN SDGs) (UN, 2015). These two developments influence each other in multiple ways, which is discussed in this chapter.

The radio spectrum continues to be the key resource for any wireless connectivity solution and presents a major control point to stakeholders. In general, spectrum management decisions aim at maximizing the value of spectrum, its efficient utilization, and benefits to the society (Beltran, 2017). The role of spectrum management is crucial

DOI: 10.1201/9781003156765-8   **125**

in defining which spectrum bands are used by the different wireless systems and by whom – important decisions that ultimately shape the market and operational models by defining which stakeholders can (re)use the precious natural resource. Given that sustainability has penetrated all aspects of our society starting from children's school-books, it cannot be excluded from spectrum management.

Spectrum decisions are all about stakeholder management (Freeman, 1984) – an important theme that yet is under-examined in telecommunications research and not properly used when regulatory bodies are making spectrum management decisions. Ultimately, the spectrum decisions have long-term impacts, which are very different for the different stakeholders who have conflicting agendas. But no one really oversees what is best for our sustainable future. Those who hold a strong position today get to significantly influence the definition of the future. Those who bring disruptive ideas in the future are not necessarily represented in that decision making. Moreover, organizations participate in multiple ecosystems simultaneously, promoting the views of not only themselves but also their partners. Some sway influence even when they are not the relevant stakeholder. These changing business ecosystems, and the simultaneous participation in multiple ecosystems, create a complex network of stakeholders with highly distinct and varying goals. A key constraint in spectrum management is how to ensure that long-term compromises are found that meet the changing needs in a sustainable way. The continuous need for more spectrum by a variety of wireless systems should be looked at from the perspective of spectrum sharing, where different systems can operate in the same band.

5G spectrum decisions have already shown growing fragmentation, about not only spectrum bands but also spectrum management approaches covering administrative allocation, market mechanisms and the unlicensed commons. Strict market structures are gradually opening for new operational models (Weber & Scuka, 2016), such as the local operators (Matinmikko-Blue, Latva-aho et al., 2017).

For the future developments towards sustainability, spectrum sharing will be a key facilitator for new business models and disruptions. When you do not have long-term stability through massive investments in spectrum, sharing-based access is highly appealing.

Spectrum sharing will be the key driver to facilitate new business models and greater inclusivity. The time for making money by creating artificial scarcity of spectrum must now transition to an era of sustainable resource management. If we fail to act soon, we will miss the target of 2030, when the UN SDGs should be achieved and the first 6G gadgets and systems will be available in the market.

7.2 Building a Joint 6G Vision

The global-scale deployment of 5G networks is ongoing. At the same time, research towards 6G has started. Finland launched the world's first 6G research program – 6G Flagship (6G Flagship) in May 2018. Since then, several 6G research programs have started in many countries and continents, bringing together academics, industry and other stakeholders.

The 8-year long 6G Flagship research effort, appointed by the Academy of Finland and led by the University of Oulu, started 6G research with a vision for 2030, where our society will be data-driven, enabled by near instant, unlimited wireless connectivity. Joint vision building for 6G with stakeholders was globally launched in March 2019 when the 6G Flagship programme organized the world's first 6G event, the 6G Wireless Summit, that gathered major telecom players to present their first visions of 6G. The event started the preparation process of a 6G White Paper with an invited group of 70 experts from around the world. The outcome – the world's first 6G White Paper (Latva-aho & Leppänen, 2019) published in September 2019, labelled 6G as "Ubiquitous Wireless Intelligence". The joint vision building reached a consensus that 6G research and development should be driven by UN SDGs that also target the year 2030. The white paper also highlights the integration of sensing, imaging and highly accurate positioning capabilities with the communication service in 6G which in turn can open a myriad of new applications in 6G. New capabilities, combined with mobility and AI/ML, can open many new application areas leading to new business opportunities in a truly digitalized society, alleviating the digital divide.

The world's first 6G White Paper (Latva-aho & Leppänen, 2019) emphasized the local operator paradigm (Zander, 2017;

Matinmikko-Blue, Latva-aho et al., 2017) where the role of local indoor networks is emphasized and different stakeholders can have their own networks, independent of mobile network operators (MNOs), through local spectrum access. Regarding the performance of 6G networks, many of the key performance indicators (KPIs) used for 5G continue to be valid also for 6G. However, there is a need to critically review the KPIs and consider new KPIs – especially towards sustainability.

Following the growing global interest towards the joint 6G vision building, the 6G Flagship programme launched another round of collaborative 6G White Paper preparation in late 2019. This time, an open call was published inviting experts to join 12 thematic groups. A set of 11 new 6G White Papers were published in June 2020 on themes that were identified based on the first white paper. A total of 250 experts from more than 100 organizations in 30 countries became contributors to the white papers. The white papers present a more detailed analysis of 6G's in terms of its linkage to the UN SDGs, business, trials for verticals, remote area connectivity, networking, machine learning, edge intelligence, trust, security and privacy, broadband connectivity, machine type communications and localization and sensing.

7.3 Connecting 6G with the UN SDGs

Following the consensus of using the UN SDGs as the key driver for 6G research and development in Latva-aho & Leppänen (2019), one of the 2020 edition 6G White Papers (Matinmikko-Blue et al., 2020) specifically addressed the linking between 6G and the UN SDGs. The white paper identified megatrends influencing the sustainable development of 6G and reviewed existing work on the broader connection between information and communications technologies (ICT) and the UN SDGs. The white paper proceeded to develop a novel linking between 6G and the UN SDGs through the existing indicators of the UN SDG framework and started the development of new 6G indicators. The white paper defined a three-fold role for 6G as (1) provider of services to help reaching the UN SDGs, (2) enabler of measuring tools for data collection to help with the reporting of indicators and (3) reinforcer of a new ecosystem to be developed in line with the UN SDGs.

The role of ICT in helping the achievement of the UN SDGs follows a three-layer approach: deployment of infrastructure and networks that form the foundation for digital economy, access and connectivity allowing people to use mobile services and enabling services and relevant content for people (GSMA, 2018; ITU, 2020). Agenda 2030 for sustainable development consist of 17 goals that cover the major global challenges. The goals are broken down to specific targets whose achievement is measured with a set of indicators. The goals address significant global challenges, such as poverty (SDG 1), hunger (SDG 2), gender equality (SDG 5) and climate change (SDG 13). The UN SDG framework with its 17 goals, 169 targets and 231 indicators presents a comprehensive action plan whose achievement can be significantly contributed to with ICT. However, out of the 231 indicators, only 7 indicators are identified as being ICT-related and address only four SDGs (SDG 4: quality education; SDG 5: gender equality; SDG 9: industry, innovation and infrastructure; and SDG 17: partnerships). These ICT-related indicators include proportion of schools with access to the Internet for pedagogical purposes, proportion of schools with access to computers for pedagogical purposes and proportion of youth/adults with ICT skills, by type of skills; proportion of individuals who own a mobile telephone; percentage of the population covered by a mobile network, broken down by technology; and fixed Internet broadband subscriptions, broken down by speed and proportion of individuals using the Internet.

In reality, there is linkage between ICT and each of the 17 SDGs. The 6G White Paper (Matinmikko-Blue et al., 2020) collected the current relation between ICT and the UN SDGs from various sources and developed a new linking between the upcoming 6G systems and the UN SDGs – both targeting the year 2030. The expert group preparing the white paper identified global megatrends, which will drive 6G research and shape the world, and looked at sustainability in a broad sense, covering political, economic, societal, technological, legal and environmental perspectives. A key observation in the white paper was that the future 6G systems, where the communication capabilities are merged with sensing, location, imaging and other capabilities, could gather a variety of data to report on the achievement of the UN SDGs on a highly local granularity level. This could respond to the challenge that nations face in how the targets defined in the UN SDG framework are being reached, noting that counter effects could also emerge. Therefore, it is important to investigate what kind of data should be collected and how and to whom it should be reported. The UN SDG framework itself is likely to evolve and during the full deployment of 6G in the 2030s, the indicator set for global challenges could be different.

In the 6G and UN SDGs white paper (Matinmikko-Blue et al., 2020) and the accompanying White Paper on Business of 6G (Yrjölä et al., 2020), the future 6G ecosystem is expected to be built around a number of new stakeholder roles and principles. Pure business-driven operations will be complemented with new societal models including community-driven networks, which will emerge depending on the regulatory environment. Another big transformation will come from the vertical industries and their public sector counterparts to whom the achievement of the UN SDGs will place significant economic constraints, and they will need to take everything the future technologies can offer to improve existing systems and processes through digitalization. This requires an early engagement of the relevant stakeholders in the process of 6G development instead of waiting for the telecommunication industry to define what 6G can bring for them, which took place in the 5G development.

7.4 6G and Spectrum Access

Spectrum regulators are in a key position to shape the future societies through their spectrum management decisions by allocating spectrum bands among different radio communication services and assigning access rights to different stakeholders. This is a complex decision-making process, where the high-level goals of maximizing the value of spectrum (Bazelon & McHenry, 2013), its efficient utilization and benefits to the society (Beltran, 2017), can be interpreted quite differently. The decision making includes different types of inputs with potentially a high level of uncertainty when predicting what impact the spectrum management decision will have in the long term. Especially, the stakeholders involved can have highly conflicting views, providing different data to back up the positions.

The role of spectrum continues to be important in the development of the next generation of mobile communication systems. Competition over the scarce spectrum resource continues to be fierce between the different wireless services. Over time, spectrum management approaches have evolved from administrative allocation (Levin, 1970) towards market-based mechanisms (Beltran, 2017; Berry et al., 2010; Hazlett, 2008; Melody, 1980; Valletti, 2001) and the unlicensed commons approach (Bazelon, 2009). Exclusive access to spectrum often is the desired model for many stakeholders. With a handful of exceptions, and despite decades of extensive research, development work and trials of spectrum sharing technologies and concepts (Guirao et al., 2017; Matinmikko-Blue et al., 2018), not much has changed in regulation.

On the one hand, the complexity of spectrum bands and access models has increased in 5G, including very different types of spectrum bands to operate in terms of carrier frequencies and bandwidths. As an example, the European pioneer bands for 5G (700 MHz, 3.5 GHz and 26 GHz) have very different characteristics leading to highly distinct deployments in terms of e.g. cell sizes and indoor/outdoor feasibility. At the same time, the variety of spectrum assignment methods used in these bands for 5G deployments varies a great deal between countries including examples of both administrative allocation and market-based mechanisms (Matinmikko-Blue et al., 2019). Additionally, 5G variants are aiming at unlicensed commons operations, further expanding the cellular networks to all known spectrum management models.

On the other hand, divergence between the spectrum assignment methods chosen for the bands to deploy 5G networks differ significantly between countries, even inside Europe. Moreover, some countries have used the very same assignment principles in all three bands, although their propagation and deployment characteristics differ drastically. Some countries have introduced new obligations in their spectrum awards decisions, while others have auctioned bands without the obligations, such as coverage requirements, that were attached to 4G spectrum. Some counties have reserved parts of the bands for local use through administrative allocation, but these bands vary from country to country. The high fragmentation between the spectrum management decisions made in the different countries about where, how and by whom 5G networks can be deployed is a new phenomenon, which places the sovereignty of the countries in deciding how to handle the spectrum access above the larger market. The same procedural approach now heavily present in 5G could also continue in 6G if best practices are not analyzed and developed properly.

Mobile communication market structures are changing (Weber & Scuka, 2016; Yrjölä et al., 2020). Existing stakeholders are taking new and different roles, and entrants are emerging (Matinmikko-Blue, Latva-aho et al., 2017). One example is local and private 5G networks that are independent of the MNOs. These developments previously discussed (Matinmikko-Blue, Latva-aho et al., 2017) through the concept of local micro operators were strongly opposed at first but have recently become accepted in several countries (Matinmikko-Blue et al., 2019). The emergence of the local 5G networks is dependent on spectrum availability, and if spectrum is in the hands of existing MNOs, the market will be fully controlled by current strong players, and there will be little incentive for change. The divergence of spectrum decision making to allow local 5G networks is high – some countries have made it happen through the availability of local spectrum licences directly without MNO involvement, while other countries rely on MNOs willingness. These spectrum management decisions significantly shape the market and can lead to missed market opportunities.

The complexity in terms of the same device operating in a wide variety of spectrum bands under different spectrum access models is also significant and end users are typically unaware of which technology is being used. The traditional split into radiocommunication services in the spectrum management domain is not consistent with the use of wireless technologies (e.g., 5G, 6G) for digitalization of the entire society. 6G combines the communication service with other services which further blurs the split. Timescales of international-level spectrum management also present a challenge in terms of reacting to the fast pace of technology development and changing user needs when the same wireless technology can be used for multiple purposes.

An ever-increasing variety of spectrum bands for mobile communications, with drastically different propagation and deployment characteristics, fragmentation of spectrum management approaches and rivalry between systems competing over spectrum access, is leading to a situation where spectrum sharing will finally become necessary. However, stakeholder management will be the bottleneck.

7.5 Time for Proper Stakeholder Management in Spectrum Management

Both sustainable development and spectrum management call for proper stakeholder engagement. While the term "stakeholder" is commonly used in many contexts, especially in spectrum management decision making by the regulators and regulatory bodies, there is very little formal background for its use. In fact, stakeholder management (Freeman, 1984) can provide a solid foundation for understanding and handling of conflicting views of the different parties. Proper spectrum management calls for understanding the stakeholder salience (Mitchell et al., 1997), which is the degree to which managers give priority to competing stakeholder claims.

In its basic form, stakeholder analysis (Freeman, 1984) consists of three steps: identification of stakeholders, stakeholder dynamics and interactions and stakeholder management actions. In the first step, stakeholders are identified in the considered context, such as the future use of a specific spectrum band like the ultra-high frequency (UHF) band in Matinmikko-Blue, Mustonen et al. (2017), noting that an organization could have highly distinct stakeholder maps depending on the context. In the second step, the roles and relations of the stakeholders' networks are investigated. This is an important step since stakeholder relationships do not occur in a vacuum of dyadic ties but rather in a network of influences. These especially business relations significantly impact the positions of the stakeholders, but they are not often seen clearly if not analyzed properly. Moreover, a single company can have highly distinct relationships with other companies in another context, which impacts their positions on e.g. spectrum assignment. It is important, therefore, to characterize the partner network of the key stakeholders carefully. This can result in new information on why stakeholders present the view that they do – it is not always their own view but that of a major customer. In the third step, the stakeholder management strategies are developed to define the actions to be taken with the stakeholders.

Although heavily studied in many other fields, stakeholder analysis (Freeman, 1984) had not been applied in spectrum management or wireless communications more generally until (Matinmikko-Blue, 2018) where a stakeholder analysis framework for sharing-based spectrum governance models to reach long-term compromises was proposed. The considered case studies, the future use of the UHF band (Matinmikko-Blue, Mustonen et al., 2017) and the Licensed Shared Access (LSA) concept (Matinmikko-Blue et al., 2014), showed examples of different stakeholders' positions on the future spectrum use. The UHF discussions showed how MNOs wished to gain access to the band, while the incumbent broadcast network operators wished to continue using the band. This gives interesting insights into the decision-making process of the national regulators or the regulatory bodies in regional and international levels. The salience of stakeholders varies – some are heard more than others, and there are big national-level differences that propagate to multi-national fora. The methods used to seek compromises between these views are often incomplete and allow the current dominant players to continue to control the discussions.

The White Paper on 6G Drivers and the UN SDGs (Matinmikko-Blue et al., 2020) introduces a preliminary action plan for engaging different stakeholders to support the achievement of the UN SDGs with the help of 6G. The work recognizes that the role of ICT is critical in meeting the UN SDGs and it is not enough to treat them separately. 6G will be a new technology generation designed fully in line with the UN SDGs and to support the achievement of the UN SDGs. Defining 6G is a matter not only for technology companies and network operators but also for the wider community that is working hard to complete the SDGs. The UN SDG framework will also need to evolve along with the technology development – current indicators do not fully reflect even today's ICT status. The white paper emphasizes that the role of ICT should be seen broadly, not only through the seven ICT-related indicators in the UN SDG framework but also through its role in helping to achieve all 17 SDGs. This can be identified through investigating how the use of new technology can contribute to the existing indicators of the UN SDG framework beyond the ICT-related indicators. Stakeholder management should be specifically addressed in the sustainable development – the technology itself is not a solution, but the use of the technology aligned with the goals is always a stakeholder's decision.

The action plan in Matinmikko-Blue et al. (2020) identifies specific roles for stakeholder groups in the joint development and evaluation effort. Governments play a key role in contributing to coverage and low cost of service for everyone through the creation of the regulatory framework and incentives to invest and operate the systems. Often more flexibility is needed to allow low-cost solutions in challenging areas that are not of business interest to operators. For the mobile communication sector, 6G is not only about developing yet another generation but also a true opportunity to contribute to sustainability at large. The role of the research community is important in facilitating stakeholder interactions and providing unbiased research results.

Spectrum management decisions should be a result of a careful and transparent stakeholder analysis process. Currently, what is called stakeholder analysis is conducted quite differently by different decision-making bodies. Stakeholder views are often collected in public consultations, but how these views are considered varies significantly. In some cases, point-to-point analyses are done addressing the stakeholders' inputs and carefully justifying the decisions taken. In other cases, stakeholder viewpoints are collected but not much is reflected in the decision making and justifications are not given. It can be observed that decisions made regarding 5G spectrum awards were dominated by the needs of those with existing strong market positions. Incumbent views on future needs were considered above all others. In instances where input from different stakeholders was considered, and where innovation, experimentation and disruption were encouraged, new practices (e.g. local licensing and new obligations that benefit consumers being included as part of the auction process) created meaningful positive change. The voices of the different stakeholders need to be heard regarding a future where sustainability must become the new norm. How do we make sure that the stakeholders developing our sustainable future can be heard? This leads to important questions – who oversees the stakeholder analysis process of spectrum assignments? Who makes sure there is room for innovation? Who looks after the interests of end users and those without a dominant position?

One thing is clear – there is not enough spectrum for every company and organization requesting it. Stakeholders have different positions, but only some stakeholders are heard in the process of today's spectrum management decisions that will impact our future. This has been particularly evident in the spectrum sharing-related discussions – those who currently control spectrum access rights are not in favour of sharing-based spectrum access models. But if sharing is a stakeholder's only opportunity to access spectrum, and the users' basic needs can be met, it becomes the most attractive option.

7.6 Conclusions

It is of utmost importance to involve the relevant stakeholders in the process of spectrum access and the sustainable development of future wireless systems. Understanding stakeholder intentions and interactions – why they say what they say must be made important. The connectivity market structure and associated stakeholder landscape are changing. What is the role of different stakeholders in envisioning the future? How will we reconcile the needs of newcomers with innovative ideas, existing players with strong market positions and authorities in charges of sustainable decisions, not to mention the citizens with growing concerns that range from legitimate to conspiratorial.

6G research is well underway globally with the Finnish 6G Flagship's start in 2018. There is a common consensus that the UN SDGs should be considered as the starting point for 6G research and development. Although they both target the same year, one should be achieved by 2030, whereas the other will only start to be deployed then, yet their interaction is important in realizing a sustainable future. The world's first 6G white paper (Latva-aho & Leppänen, 2019) built a connection between 6G and the UN SDGs. The second white paper (Matinmikko-Blue et al., 2020) specifically stressed the urgency of investigating the role of ICT in helping the achievement of the UN SDGs through providing services, collecting data and developing the telecommunication systems in accordance with the UN SDGs. In fact, the UN SDGs and related regional and national-level policy goals have already started to impact various verticals, placing increasing requirements on renewing their operations to become sustainable. The achievement of the sustainability goals is highly dependent on the use of ICT solutions which themselves also need to be sustainable.

When it comes to spectrum management for future systems, spectrum sharing is the way towards a sustainable future. To make it happen, stakeholder management needs to undergo a major transformation. The variety of bands and fragmentation will keep increasing as we head towards 6G – the role of spectrum sharing is critical in the landscape of changing market structures and stakeholder needs.

References

6G Flagship. www.6gflagship.com.

C. Bazelon. (2009). Licensed or unlicensed: The economic considerations in incremental spectrum allocations. *IEEE Communications Magazine*, vol. 47, no. 3, pp. 110–116.

C. Bazelon and G. McHenry. (2013). Spectrum value. *Telecommunications Policy*, vol. 37, pp. 737–747.

F. Beltran. (2017). Accelerating the introduction of spectrum sharing using market-based mechanisms. *IEEE Communications Standards Magazine*, vol. 1, no. 3, pp. 66–72.

R. Berry, M. L. Honig, and R. Vohra. (2010). Spectrum markets: Motivation, challenges, and implications. *IEEE Communications Magazine*, vol. 48, no. 11, pp. 146–155.

R. E. Freeman. (1984). *Strategic management: A stakeholder approach*. Boston: Pitman Publishing Inc.

GSMA. (2018). *Mobile industry impact report: Sustainable development goals*. www.gsmaintelligence.com/research/?file=ecf0a523bfb1c9841147a335 cac9f6a7&downloadM.

D. P. Guirao, A. Wilzeck, A. Schmidt, K. Septinus, and C. Thein. (2017). Locally and temporary shared spectrum as opportunity for vertical sectors in 5G. *IEEE Network*, vol. 31, no. 6, pp. 24–31.

T. W. Hazlett. (2008). Optimal abolition of FCC spectrum allocation. *Journal of Economic Perspectives*, vol. 22, no. 1, 103–128.

ITU. (2020). *Connect 2030 Agenda*. International Telecommunication Union. https://itu.foleon.com/itu/connect-2030-agenda/home/.

M. Latva-aho and K. Leppänen (Eds.). (2019). Key drivers and research challenges for 6G ubiquitous wireless intelligence. [White paper]. (6G Research Visions, No. 1). University of Oulu, Finland. http://urn.fi/urn: isbn:9789526223544.

H. J. Levin. (1970). Spectrum allocation without market. *The American Economic Review*, vol. 60, no. 2, pp. 209–218.

M. Matinmikko-Blue. (2018). *Stakeholder analysis for the development of sharing-based spectrum governance models*. Doctoral thesis in Industrial Engineering and Management, University of Oulu, Finland. http://jultika.oulu.fi/files/isbn9789526220512.pdf.

M. Matinmikko-Blue, M. Latva-aho, P. Ahokangas, and V. Seppänen. (2018). On regulations for 5G: Micro licensing for locally operated networks. *Telecommunications Policy*, vol. 42, no. 8, pp. 622–635.

M. Matinmikko-Blue, M. Latva-aho, P. Ahokangas, S. Yrjölä, and T. Koivumäki. (2017). Micro operators to boost local service delivery in 5G. *Wireless Personal Communications*, vol. 95, no. 1, pp. 69–82, July.

M. Matinmikko-Blue, M. Mustonen, and H, Haapasalo. (2017). Stakeholder analysis for future use of the ultra-high frequency (UHF) band. *International Journal of Technology, Policy and Management*, vol. 17, pp. 159–183.

M. Matinmikko-Blue, H. Okkonen, M. Palola, S. Yrjölä, P. Ahokangas, and M. Mustonen. (2014). Spectrum sharing using Licensed Shared Access: The concept and its workflow for LTE-Advanced networks. *IEEE Wireless Communications*, vol. 21, pp. 72–79.

M. Matinmikko-Blue, S. Yrjölä, V. Seppänen, P. Ahokangas, H. Hämmäinen, and M. Latva-Aho. (2019). Analysis of spectrum valuation elements for local 5G Networks: Case study of 3.5-GHz band. *IEEE Transactions on Cognitive Communications and Networking*, vol. 5, no. 3, pp. 741–753, September.

M. Matinmikko-Blue, et al. (Eds.). (2020). White Paper on 6G Drivers and the UN SDGs. [White paper]. (6G Research Visions, No. 2). University of Oulu. http://urn.fi/urn:isbn:978952622669.

W. H. Melody. (1980). Radio spectrum allocation: Role of the market. *The American Economic Review*, vol. 70, no. 2, pp. 393–397.

R. K. Mitchell, B. R. Agle, and D. J. Wood. (1997). Toward a theory of stakeholder identification and salience: Defining the principle of who and what really counts. *Academy of Management Review*, vol. 22, pp. 853–886.

UN. (2015). *The 2030 Agenda for Sustainable Development*. United Nations. https://sdgs.un.org/2030agenda.

T. M. Valletti. (2001). Spectrum trading. *Telecommunications Policy*, vol. 25, no. 10–11, pp. 655–670.

A. Weber and D. Scuka. (2016). Operators at crossroads: Market protection or innovation? *Telecommunications Policy*, vol. 40, no. 4, pp. 368–377.

S. Yrjölä, P. Ahokangas, and M. Matinmikko-Blue (Eds.). (2020). White paper on business of 6G. [White paper]. (6G Research Visions, No. 3). University of Oulu. http://urn.fi/urn:isbn:978952622676.

J. Zander. (2017). Beyond the ultra-dense barrier: Paradigm shifts on the road beyond 1000x wireless capacity. *IEEE Wireless Communications*, vol. 24, no. 3, pp. 96–102, January.

8

HOW CAN SPECTRUM POLICY ADDRESS CLIMATE CHANGE?

MANUEL R. MARTI

Contents

8.1 Introduction

Digital tools and services have grown rapidly over the last few decades and have become an integral part of our lives. One could hardly imagine a life without smartphones, apps, route planners or an endless selection of music and movies at our fingertips.

These trends in the information and communications technology (ICT) field are turning the sector into a significant contributor to global greenhouse gas (GHG) emissions. The amount of data we are sending and receiving comes with a burden – energy consumption is on the rise and so is the sector's carbon footprint.

Across the entire sector – including data centres, fixed and mobile networks and devices – ICT accounts for about 2%[1] of global carbon emissions. Recent estimates,[2] however, suggest that figure could jump to 14% of worldwide emissions by 2040.

DOI: 10.1201/9781003156765-9

Paradoxically, ICT is at the heart of many solutions[3] that reduce energy demand and GHG emissions. By increasing connectivity and efficiency, ICT services could help to avoid emissions across all sectors. A frequently cited example is that video conferencing and telecommuting[4] could reduce air and/or road travel, which causes high amounts of GHG emissions due to fossil fuel consumption.

ICT-enabled improvements have historically gone hand in hand with increases in energy consumption and GHG emissions both within the ICT sector and in the wider economy, however. Data centres appear to be one of the fastest growing GHG-emitting entities and the largest polluter[5] within ICT, followed by mobile and fixed networks.

More specifically, it is estimated[6] that total GHG emissions from mobile and fixed networks (including both manufacturing and use) amount to about 0.4% of global GHG emissions. The vast majority of these emissions are associated with the electricity consumption during use. As of today, mobile networks make up for approximately[7] 0.6% of the global electricity consumption.

The mobile sector's share of electricity is non-trivial, which is challenging for mobile operators because rising electricity costs form a significant part[8] of the operational expenditure (OPEX). The introduction of each new generation of mobile networks – similarly to general ICT developments – has historically increased the total energy consumption. Intake does not increase with the standard itself but rather depends on how networks are deployed and used. Extensive and congested networks have consequently led to higher greenhouse emissions. According to recent studies,[9] 90% of telcos believe that the rollout of the fifth generation of mobile networks (5G) will result in even higher electricity costs.

These worrying trends come at a time when climate change could not have a higher profile in the international community. ICT environmental footprint and potential have been long studied over the last two decades, although the role of spectrum management in environmental sustainability remains an open question at this point. There is, however, a growing willingness across industry, academia and legislators to explore how spectrum policies could help the fight against climate change.

Is the impact of spectrum policy necessarily a small part of a sector-wide environmental approach? Is additional regulation, such as enforced carbon caps within spectrum licences, needed to achieve sustainability in the wireless sector? Or are the commercial benefits of reducing energy consumption sufficient incentive? Those are the questions that European institutions, in particular, and the wider International Telecommunication Union (ITU) community are currently grappling with.

This chapter explores the relationship between climate change and radio spectrum management, by analysing all the different dimensions in which spectrum regulation could be directly or indirectly involved. Specifically, it presents an overview of the ICT industry current energy consumption and carbon emissions (Section 8.2) and looks at the increase in interest from policymakers to tackle this concerning issue (Section 8.3). The chapter also examines potential savings that can be obtained from greater 5G use (Section 8.4) and explores the role of regulators while questioning how self-regulation could minimise the sector's environmental footprint (Section 8.5). Finally, it presents some non-conventional examples, such as satellite protection e-waste and 6G, where an alternative approach to frequency management could achieve possible carbon savings.

8.2 ICT Carbon Footprint: The Jevons Paradox

Given the relevance of the topic, there is surprisingly little literature analysing the environmental impact of ICT. There also seems to be a lack of agreement about which technologies should be included in calculations of ICT's GHG emissions.

For such reasons, different studies have come up with slightly different figures. The most optimistic academics[10] suggests that the ICT sector itself contributes around 1.4% of annual global carbon emissions, while more gloomy views[11] claim it accounts for 3.5% of the total emissions. According to the ITU's SMART 2020 report,[12] the carbon footprint of the ICT industry amounts to 2% of global emissions, while the European Commission[13] puts the whole sector's share of global GHG emissions just over 2%.

In light of constant technological progress, increased efficiency and changing consumer habits, it is difficult to estimate exactly how large the global CO_2 emissions are, although most studies – especially those carried by the ITU and GeSI[14] – seem to focus on ICT's rather positive effect in other markets.

The effect of ICT depends on the balance of impacts it has both through its own emissions and the effects it has on the wider economy, but it is often argued that there is a huge potential for these technologies to mitigate GHG emissions in other sectors, such as transport, manufacturing, construction and/or agriculture. Common examples include video conferencing technologies or online shopping which could reduce the need for physical travel, or IoT and sensing technologies which could improve efficiencies in the supply chain or the use of water in agriculture.

While these claims may be true and ICT has the potential to decrease the carbon footprint of other industries, these continuous efficiency gains have also gone hand in hand with an over increase in use, which translates into an increase[15] in energy consumption, as well as GHG emissions. ICT's energy consumption and, therefore, its carbon emissions have outpaced efficiencies and gains experienced within the ICT sector itself and other industries year after year.

One wouldn't be totally wrong to assume that ICT's historical environmental pattern fits with the rebound effect described by Jevons Paradox – in which efforts to improve energy efficiency can more than negate any environmental gains. Jevons Paradox[16] refers to a situation in which an efficiency improvement leads to an even greater proportionate increase in total demand, with the result that resource requirement goes up rather than down, as is often taken for granted.

The question is not whether ICT has energy and environmental costs, but whether those costs can be mitigated, and whether they are offset by the energy and environmental benefits.

An illustrative example[17] of such paradox is the fact that electric trains are hugely more efficient than steam trains, although the environmental impact of land transport has continuously risen over time.

8.3 Increasing Role of Spectrum Policy in Carbon Reductions

Over the last few years, climate change has become a priority for policymakers and governance structures at all levels, including in industry, governments and academia.

Carbon reduction targets have been set at global, regional and many national levels to tackle climate change. The Paris Agreement was the first universal, legally binding global climate change agreement. Adopted in December 2015, close to 190 countries agreed the long-term target to limit the increase in global average temperatures to 2°C above pre-industrial levels.

Accordingly, this has placed the spotlight on the ICT sector. The UN-agency ITU has long studied the industry's environmental footprint, but recent years saw the international organisation place intense focus[18] on sustainability and climate protection programmes.

In February 2020, the ICT industry became the first to develop sectoral targets approved[19] by the ITU, which helps companies set emissions targets in line with the UN Framework Convention on Climate Change, agreed in Paris. Twenty-nine telecom companies representing 30% of mobile connections worldwide have committed to reduce emissions by 45% by 2030, from 2020 levels.

For the first time, there is also an increasing awareness of the potential role of spectrum. Like all sectors, spectrum policy will have to do its part to address global climate change, with the ultimate goal of achieving net-zero emissions. A sense of urgency has grown markedly over the last two years as the potential of spectrum to reduce GHG emissions is increasingly acknowledged as relevant in intergovernmental policy documents.[20]

Europe appears to be leading the way in implementation of, and experimentation with, climate policy. The Commission's Green Deal, signed in 2019, aims to cut net emissions of greenhouse gases to zero by 2050 and represents a serious shift in European policy.[21] Although it does not provide a detailed roadmap, it creates a new agenda for the bloc that will shape its development – including spectrum policy – in the years to come.

Radio spectrum management features prominently in policymaking around the climate. In its digital strategy published last year, the European Union revealed that its Radio Spectrum Policy Programme (RSPP) – which will define the region's spectrum direction for 2025–2030 – will have an environmental dimension.

The European Commission's spectrum advisory group, the RSPG, emphasised in its draft opinion[22] on the review of the spectrum programme (RSPP) that countries should examine how different spectrum initiatives could help the Commission meet its environmental targets.

It noted[23] that the European Commission in cooperation with EU member states should ensure that adequate spectrum is made available under harmonised conditions to support EU initiatives to combat climate change and improve energy saving.

The RSPG seems to have acknowledged the limitations of current assessments of GHG emissions as it called for new methodologies to assess the impact of wireless technologies on climate change.

In a similar vein, another Commission's document published in 2020 suggests that additional regulation may be needed to achieve sustainability in the telecoms sector. A 5G Recommendation[24] urges countries to consider all possible means of counteracting the rise of climate change and claims that incentives to deploy networks with, for example, a reduced carbon footprint can contribute to the sustainability of the sector and to climate change mitigation and adaptation.

It seems, however, that the European Union's clear policy focus is to mitigate the impacts associated with rising consumption, specifically through improved efficiency and renewable energy, instead of seeking to directly cut this consumption trend. According to the same policy paper,[25] member states are encouraged to develop best practices to incentivise the deployment of electronic communications networks with a reduced environmental footprint, particularly with respect to energy use and related GHG emissions, including (1) the criteria for assessing the environmental sustainability of future networks and (2) the incentives provided to operators to deploy environmentally sustainable networks.

In conclusion, the spectrum world has just started making its own contribution to emissions reductions, but this matter appears to be ranked very high within policy circles. Brussels officials[26] have personally recognised that the footprint of the electronic communications sector is increasing and questioned to what extent spectrum could help curb this trend.

In the following sections, this chapter will first look at the environmental impact of the next generation of mobile networks, then it will look at sector's self-regulation initiatives and discuss the role of tele-coms regulators and finally it will present some innovative examples where spectrum policies could do its part.

8.4 Will 5G Be More Energy Efficient than Its Predecessors?

The introduction of each new generation of mobile networks has historically increased total energy consumption – intake does not increase with the standard itself but rather depends on how networks are deployed and used. Extensive and congested networks have con-sequently led to higher greenhouse emissions. This growth has led to concerns[27] that the next generation of mobile networks will consume an increasing share of electricity, thereby accelerating climate change.

Vendors and mobile operators recognise that 5G could have neg-ative effects on the environment because of its energy use and the impacts of manufacturing new infrastructure and a multitude of new devices. In doing so, private companies have set up aggressive tar-gets for emissions reduction in their sustainability strategies. How-ever, most of these plans[28] rely on 5G's dual role in mitigating climate change. Firstly, it can address sector emissions through energy effi-ciency and renewables uptake. Secondly, it has the potential to deliver emissions reductions across the wider economy.

A lot of work within the ITU[29] and other standardisation and engineering organisations, such as ETSI or IEEE, focuses on energy efficiency metrics and measurement methods for radio networks and devices. From a standards point of view, 5G is designed to be more energy efficient than previous generations. New energy-saving soft-ware, innovative cooling techniques as well as advanced network planning and deployment will allow 5G to be more energy efficient. However, that only applies on a per-bit basis: 5G will be able to send more data bits per kilowatt of energy than any previous generation of wireless technology.

Efficiency efforts have been further shaped by a number of industry-specific factors rooted in countering rising network costs. For exam-ple, Ericsson notes in its 'Breaking the Energy Curve' report[30] that 5G is 90% more energy efficient compared with 4G, while Huawei claims that GHG emissions per connection can drop as much as 80% by 2025 – as long as 5G is operational by then.

Improving energy efficiency is only one dimension of the challenge. Another is to shift from carbon to non-carbon energy sources. As of today, the ICT sector[31] is the world's largest purchaser of renewable electricity, and many companies are meeting their clean energy pledges through power purchase agreements (PPAs), in which they purchase renewable energy credits equal to their energy demand. However, matching energy consumption with clean energy in real time is more difficult, as variable generation from wind and solar may not match the electricity demands of networks or data centres.

5G, which began deployment in 2019, can deliver not only enhanced broadband for cell phones, more reliable communication, but also enhanced machine-to-machine communication. These characteristics appear to offer great potential for reducing the energy requirements and carbon footprint of other sectors by (1) improving and optimising equipment and infrastructures (e.g., a smart thermostat optimises heating or a smart irrigation system could help save water) and (2) substituting travelling and the transport of goods through new applications running over high-bandwidth (video conferencing can be a substitute for air travel and physical meetings).

The mobile industry often highlights[32] such traits when addressing climate concerns – how 5G-induced substitution and optimisation effects can lead to a decrease in GHG emissions. However, it fails to accentuate that the sector's overall energy consumption and carbon emissions remain on the rise. There are risks that counteract the GHG abatement potential of the 5G-supported use cases. For example, rebound effects can compensate, if not overcompensate, for the expected GHG reductions. It is Jevons' Paradox: an increase in energy efficiency can go hand in hand with a hike in carbon emissions.

As highlighted earlier, ICT and mobile networks have become dramatically more efficient and have consequently delivered wide-ranging efficiency and productivity improvements across the wider economy; however, ICT and mobile's footprint has risen to account for a significant proportion of the overall emissions, and global emissions have also increased continuously.

8.5 The Regulatory Dilemma: Self-Regulation or Enforcement

The push to be more green appears to be in full swing across the tele-coms sector. In the last few years, some of the world's largest mobile operators and vendors have announced a multitude of climate pledges, environmental goals and other social responsibility initiatives. BT, for example, said[33] in its 2019 'sustainability report' that it would like to be carbon neutral by 2045. Telefonica has also set itself a sustain-ability challenge, aiming to become a net-zero mobile network in its top four operating markets within this decade. Verizon and Vodafone have set targets to reach net zero by 2040, with both mobile carriers aiming for 50% reductions in electricity usage by 2025.

Several big ICT companies also formed a coalition[34] to boost their efforts to tackle climate change. Vodafone, Deutsche Telekom, Orange, Nokia and Ericsson were among those to sign a declaration and pledge to eliminate their contributions to carbon emissions by 2040.

The cellular industry was also the world's first, in 2016, to commit to achieving the UN Sustainable Development Goals (SDGs), which included setting an industry goal of net-zero emissions by 2050. As previously mentioned in Section 8.2, the ITU, in collaboration with GSMA, GeSI and SBTi, released a non-binding standard in 2020 that aims to reduce ICT's GHG emissions by 45% by 2030, and net zero by 2050, in line with the Paris agreement.

So far, all these carbon initiatives have been voluntary and self-regulatory. These, however, have a number of persistent barriers, including lack of transparency, or/and different methodologies. Some assert that if ICT were to be subjected to well-resourced and inde-pendent scrutiny GHG emissions would be lower. More voices are now calling for policy-enforced carbon caps on global emissions, or carbon pricing, as ICT carbon emissions continue to scale up. This will entail a completely new regulatory framework for telecommuni-cations – including spectrum policy.

A proposed scenario[35] would see future spectrum policy having a built-in energy or CO_2 function, such as a licence condition for new awards that relates to the energy used.

In other words, regulatory authorities may want to include GHG emission targets in their licensing regimes in the same way that they apply coverage obligations. That means every time that a new mobile carrier or other spectrum user (vertical markets) secures a spectrum licence they would have to commit to very specific targets.

These licence prerequisites would not necessarily have to be obligations, they could be incentives instead as it is proposed in the European Commission's recommendation.[36] These might include fast-track permit granting procedures or reduced permit and access fees for networks which meet certain environmental criteria. Even if environmental protection were to be added to regulators' list of duties, it seems unlikely that the regulators policy approach would change fundamentally.

Other regulatory proposals could encourage more efficient practices, including compression technologies, greater use of Wi-Fi, network sharing and the use of renewable sources.

Studies point out that data traffic is a driver in ICT growth and emissions, and video streaming is a particular data-intensive application. In 2020, Netflix agreed[37] with EU regulators to reduce their traffic and ease the load on the network, allowing network provision for homeworkers during the Covid-19 pandemic. Similarly, regulators could ensure that people watch video at a resolution that is sensible for the environment. Experts[38] believe that if one manages to compress high definition to standard definition, one can significantly reduce the data rates. Moreover, a greater use of Wi-Fi could technically reduce carbon emissions because it has much lower energy requirements than cellular networks due to its shorter range.

With the telecoms community becoming increasingly aware of its environmental footprint, the green agenda is playing a more significant role in decision-making. Self-regulation and public pressure may be enough to get more ICT companies to become carbon-neutral. However, there is an increasing pressure for regulators to take action because some believe that without carbon-enforced policies the sector may never become sustainable.

8.6 Alternative Dimensions: Satellite Protection, E-Waste and 6G

Satellite and other technologies have long been used to forecast weather, document environmental changes and measure sea levels, atmospheric gases and the planet's changing temperature, among other factors. These have become an essential element in mitigating the effects of climate change as predictions could act as early warning systems for extreme weather events. These satellites (including EU's Galileo, GMEs, scientific services, EESS) are governed by the ITU and new decisions regarding spectrum needs or additional protection from other services, such as 5G, are dictated at the World Radiocommunication Conferences (WRCs). This clearly shows an additional environmental dimension to spectrum regulation which is often not discussed.

In fact, the last conference (WRC-19) saw extensive and contentious policy discussions when it allocated new mmWave frequencies for 5G services. The issue was that some of the frequencies considered for 5G (26 GHz) were next to those frequencies used by satellites that gather weather information. Having this proximity means that signals could interfere with one another and consequently impact weather observations.

At the end, the conference decided on a time delay approach to the limits imposed to protect passive weather sensors at 23.6–24 GHz. According to the agreement, mobile base stations operating in the band would not be able to leak more than –33 dBW over 200 MHz into the passive band until 2027, the date of a future WRC. After WRC-27, this limit would be tightened to –39 dBW/200 MHz for new deployments only. Weather forecasters, however, remain concerned[39] that the outcome of the conference may still have an adverse impact on future Earth observation satellite systems. The risk is that 5G could roll out more quickly than initially anticipated, creating an unregulated increase in interference in the 24 GHz meteorological satellite observation frequency radio spectrum band.

One may wonder if the outcome of the conference would have been any different if the spectrum managers who make decisions on radio-frequency allocations at WRCs had a greener agenda. European spectrum managers[40] have subsequently gone further and said they will implement the tighter limits at the beginning of 2024, rather than 2027. They will also prevent any high-density deployment in 22–23.6 GHz. European countries had advocated a limit of –42 dBW/200 MHz in preparations for the conference. Similarly, the RSPG document[41] highlighted that the European Commission alongside member states should support technologies contributing to climate change monitoring/climate protection aspects. It notes that countries ensure long-term spectrum availability and protection for radio systems supporting them and cooperate as necessary in order to assess and solve radio interference into these services.

Another noteworthy policy area in which the relevance of spectrum policy is often underestimated is e-waste. Devices often carry a heavy environmental footprint, as they do contribute to carbon emissions not only while manufacturing but also at disposal. According to statistics[42] from the UN, an estimated 50 million tons of electronic waste are generated worldwide each year. The trend is on the rise, as e-waste is recognised by the World Economic Forum[43] as the fastest growing category of waste. The European Union[44] is planning to put forward a 'Circular Electronics Initiative' by the end of 2021 to improve the lifespan, repairability and recyclability of ICT products. This initiative would help decrease the embodied carbon of ICT, but the question is whether spectrum policy could contribute to reduce e-waste.

As the total number of handsets[45] continues to increase, one could argue that trend may be partially driven by certain spectrum policies. The mobile industry has continued to pursue new frequencies as they developed new generations of mobile networks. 5G, for instance, put millimetre wave (mmWave) frequencies in the spotlight, and now as people start looking forward to what 6G might hold and how to get there, terahertz spectrum[46] (0.1 THz–3 THz) seems the most plausible route.

This hunt for new frequencies, however, has its implications: some of the sector's carbon emissions will be due in part to the fact that consumers will need to update their mobile phones in order to take full advantage of the next generation of mobile phones. Some studies[47] suggest that a smartphone can produce up to 45 kg of CO_2 during its entire lifetime, with most of it coming from the production phase.

So could new spectrum policy target this problem? Could regulators incentivise e-waste recycling programmes? Should regulators and private companies focus on re-farming or re-organising the frequencies that have already been allocated? The European Environmental

Bureau (EEB) claims[48] extending the life of smartphones and other electronics by just one year would be the equivalent of taking two million cars off the road, in terms of CO_2 emissions.

Looking to the future, there may be some solid reasons to be optimistic about the future of ICT. Researchers and academics have set out sustainability as a pillar upon which to build their vision of 6G. The UN's SDGs – which include climate action, sustainable cities and communities, responsible consumption and production or the right for everyone to access affordable, reliable, sustainable and modern energy – seem to be the key principles that underpin every part of ongoing research and that seems to be where much of 6G's potential lies.

Previous and current generations of mobile networks have not excelled in this field and much more could be done to reach some environmental targets. Future technologies, such as 6G, are likely to have explicit design goals related to efficiency and sustainability, both in terms of radio network waveforms and ultimately in the form of spectrum. One of the main objectives of 6G research programmes is to assess how the new technology could help achieve the UN's global goals (SDGs).

Researchers at the University of Oulu[49] identify 6G as a three-fold facilitator for the SDGs. They think it will provide services to help steer and support countries' efforts towards the goals. It will also enable data collection for measuring progress towards the goals. Finally, it will reinforce new ecosystems based on 6G technology enablers.

So while optimism may be as scarce as spectrum in 2021, at least 6G could provide some hope.

8.7 Conclusion

As the rapid growth of the ICT industry is expected to continue, the entire sector – including governments, regulators, private firms and academic institutions – has a great responsibility to ensure that what it is deploying is sustainable. So far, the sector has failed to comply with such duty. Efficiency and energy gains achieved by deployment of ICT solutions have not been compensated by a reduction in the sector's environmental footprint – energy consumption and carbon emissions have increased on a yearly basis.

It now seems we are at a tipping point as more ICT companies have initiated sustainability and climate protection programmes. These moves reflect a growing trend in the field, as more and more institutions transition towards sustainable products and practices. However, some claim these self-regulatory actions may not be enough to overturn the sector's worrying carbon emissions trend. Thus, there is a growing willingness to explore how spectrum policies could help the fight against climate change.

It is an area where policy is not yet fully formed, although when the topic is discussed in policy circles a lot of attention is placed on the value of next-generation telecommunications in delivering on the UN's long-term sustainable objectives. But spectrum managers could do more to enable carbon reduction, besides awarding spectrum for 5G networks. For example, enforced carbon caps within spectrum licences, or alternative incentives for the rollout of greener networks could be a potential solution. Similarly, should legislators start prioritising the long-term spectrum availability and protection of weather satellites which contribute to climate change monitoring and climate protection over other technologies?

Looking ahead, there are reasons to be optimistic about the future of ICT because researchers are placing sustainability as a pillar upon which to build their vision of future technologies, including 6G. However, one would expect regulators to start questioning if moving up in frequency ranges with every new generation of mobile networks is affordable for the environment – or at least until more device buy-back or recycling plans are in place.

As consumers around the world now move to 5G phones, many older handsets will soon be discarded, adding to the growing footprint e-waste has today. The figures[50] are alarming: a 1-year lifetime extension of all smartphones in the EU would save 2.1 million tonnes (Mt) of CO_2 per year by 2030. That is the equivalent of taking over a million cars off the roads. If one increases the lifespan of handsets by three years, one would save around 4.3 $MtCO_2$, while a 5-year extension would correspond to about 5.5 $MtCO_2$. Long-term effects of ICT on the environment are unknown, but one would expect the overall sector to step up and do more against the current and growing challenges.

Notes

1. As explained later in the paper, there is little consensus on the environmental impact of ICT. This figure was taken from ITU's Summary of SMART 2020 Report. www.itu.int/md/T05-FG.ICT-C-0004/en.
2. L. Belkhir and A. Elmeligi, 2018. *Assessing ICT Global Emissions Footprint: Trends to 2040 & Recommendations.* www.electronicsilent spring.com/wp-content/uploads/2015/02/ICT-Global-Emissions-Foot print-Online-version.pdf.
3. ITU and GeSI, 2011. *Using ICTs to Tackle Climate Change.* www.itu.int/dms_pub/itu-t/oth/0B/11/T0B1100000A3301PDFE.pdf.
4. Ibid.
5. Belkhir and Elmeligi, *Assessing ICT Global Emissions Footprint.*
6. GSMA, 2019. *Energy Efficiency: An Overview.* www.gsma.com/future networks/wiki/energy-efficiency-2/.
7. Ibid.
8. Chatzimichail, 2014. *How to Cut the Electric Bill in Mobile Access Networks: A Mobile Operator's Perspective.* www.diva-portal.org/smash/get/diva2:872755/FULLTEXT01.pdf.
9. Marti, 2019. Telecoms Sector Is Slowly Starting to Address Climate Change. *Policy Tracker.* www.policytracker.com/telcoms-sector-is-slowly-starting-to-address-climate-change/.
10. J. Malmodin and D. Lundén, 2018. *The Energy and Carbon Footprint of the Global ICT and E&M Sectors 2010–2015.* Sweden: Ericsson Research.
11. Belkhir and Elmeligi, *Assessing ICT Global Emissions Footprint.*
12. ITU. *Summary of SMART 2020 Report.* www.itu.int/md/T05-FG.ICT-C-0004/en.
13. European Commission, 2020. *Supporting the Green Transition.* https://ec.europa.eu/commission/presscorner/detail/en/fs_20_281 – accessed March 2020.
14. GeSI, 2020. Smarter2020: The Role of ICT in Driving a Sustainable Future.
15. N. Jones, 2018. How to Stop Data Centres from Gobbling Up the World's Electricity. *Nature.* www.nature.com/articles/d41586-018-06610-y.
16. D. Owens, 2010. The Efficiency Dilemma. *The New Yorker.* www.new yorker.com/magazine/2010/12/20/the-efficiency-dilemma.
17. M. Berners-Lee, D.C. Howard, J. Moss, K. Kaivanto, and W. Scott, 2011. Greenhouse Gas Footprinting for Small Businesses – The Use of Input – Output Data. *Science of the Total Environment,* 409(5).
18. ITU Activities & Sustainable Development Goals, 2020. www.itu.int/en/action/environment-and-climate-change/Pages/ITU-in-the-UN-Environmental-Agenda.aspx.

19. ITU, 2020. *ICT Industry to Reduce Greenhouse Gas Emissions by 45 Per Cent by 2030.* www.itu.int/en/mediacentre/Pages/PR04-2020-ICT-industry-to-reduce-greenhouse-gas-emissions-by-45-percent-by-2030.aspx#:~:text=A%20new%20ITU%20standard%20highlights,cent%20from%202020%20to%202030.

20. European Commission, 2020. *RSPG Opinion on a Radio Spectrum Policy Programme (RSPP).* https://rspg-spectrum.eu/wp-content/uploads/2021/02/RSPG21-014final_Draft_RSPG_Opinion_on_RSPP.pdf.

21. ECIPE, 2021. *The EU Green Deal and Its Industrial and Political Significance.* https://ecipe.org/publications/eu-green-deal/.

22. European Commission, *RSPG Opinion on a Radio Spectrum Policy Programme (RSPP).*

23. Ibid.

24. European Commission, 2020, *Commission Recommendation on a Common Union Toolbox for Reducing the Cost of Deploying Very High Capacity Networks and Ensuring Timely and Investment-Friendly Access to 5G Radio Spectrum.* https://ec.europa.eu/digital-single-market/en/news/commission-recommendation-common-union-toolbox-reducing-cost-deploying-very-high-capacity.

25. Ibid.

26. Marti, 2019. RSPG Maps Out Future Agenda. *Policy Tracker.* www.policytracker.com/rspg-maps-out-future-agenda/.

27. Curran, 2020. *What Will 5G Mean for the Environment?* University of Washington. https://jsis.washington.edu/news/what-will-5g-mean-for-the-environment/#_ftn16.

28. Analysys Mason, 2020. *Green 5G: Building a Sustainable World.* www.analysysmason.com/contentassets/bb742b22fb434cf8a055291c20331dfe/analysys_mason_green_5g_sustainability_jul2020_rma18_rdns0.pdf.

29. ITU ITU-T SG5.

30. Ericsson, 2020. *Breaking the Energy Curve: How to Roll Out 5G Without Increasing Energy Consumption.* www.ericsson.com/en/news/2020/3/breaking-the-energy-curve#:~:text=%27Breaking%20the%20energy%20curve%27%20is,carbon%20emissions%20in%20mobile%20networks.

31. Ibid.

32. GSMA, *Energy Efficiency: An Overview.*

33. BT, 2019. *Impact and Sustainability Report 2019/2020.* www.bt.com/about/digital-impact-and-sustainability/our-report.

34. ETNO, 2021. *Joint Statement by ETNO and the GSMA on the European Green Digital Coalition.* www.etno.eu/news/all-news/702-telcos-egdc.html.

35. Marti, 2020. Mobile Industry Pledges to Reduce its Carbon Footprint. *Policy Tracker.* www.policytracker.com/mobile-industry-pledges-to-become-more-sustainabile/.

36. European Commission, *Commission Recommendation.*

37. Sweney, 2020. *Netflix to Slow Europe transmissions to Avoid Broadband Overload.* www.theguardian.com/media/2020/mar/19/netflix-to-slow-europetransmissions-to-avoid-broadband-overload – accessed March 2020.
38. Marti, 2021. Does Spectrum Policy have an Environmental Dimension? *Policy Tracker.* www.policytracker.com/does-spectrum-policy-have-an-environmental-dimension/.
39. Standeford, 2019. WRC-19 Outcomes Please Amateur Radio but Anger Weather Satellite Users. *Policy Tracker.* www.policytracker.com/wrc-19-reaction-from-amateur-radio-weather-satellites/.
40. Marti, 2020. Europe Chooses New Transition Date for 26 GHz Band. *Policy Tracker.* www.policytracker.com/europe-chooses-new-transition-date-for-the-26-ghz-band/.
41. European Commission, *RSPG Opinion on a Radio Spectrum Policy Programme (RSPP).*
42. UN, 2019. *Time to Seize Opportunity, Tackle Challenge of e-Waste.* www.unep.org/news-and-stories/press-release/un-report-time-seize-opportunity-tackle-challenge-e-waste.
43. World Economic Forum, 2019. *A New Circular Vision for Electronics.* Time for a Global Reboot. Platform for Accelerating the Circular Economy (PACE). https://www3.weforum.org/docs/WEF_A_New_Circular_Vision_for_Electronics. pdf – accessed March 2020.
44. European Commission, 2020d. *Changing How We Produce and Consume: New Circular Economy Action Plan Shows the Way to a Climate-Neutral, Competitive Economy of Empowered Consumers.* https://ec.europa.eu/commission/presscorner/detail/en/ip_20_420 – accessed March 2020.
45. Cisco, 2020. *Cisco Annual Internet Report (2018–2023).* www.cisco.com/c/en/us/solutions/collateral/executiveperspectives/annual-internet-report/white-paper-c11-741490.pdf.
46. Marti, 2019. Will THz Spectrum Ever Be Usable? *Policy Tracker.* www.policytracker.com/will-thz-spectrum-ever-be-usable/.
47. Ercan, 2013. *Global Warming Potential of a Smartphone: Using Life Cycle Assessment Methodology.* www.semanticscholar.org/paper/Global-Warming-Potential-of-a-Smartphone-%3A-Using-Ercan/0db4e4c6396ec9ce3924969f70ec5dad016f3c7e?p2df.
48. EEB, 2019. *Coolproducts Don't Cost the Earth – Full Report.* www.eeb.org/coolproducts-report.
49. Matinmikko-Blue et al., 2020. *White Paper on 6G Drivers and UN Sustainable Development Goals.* www.6gchannel.com/items/6g-white-paper-6g-drivers-un-sdgs/.
50. EEB, *Coolproducts Don't Cost the Earth.*

9

ARTIFICIAL INTELLIGENCE IN RADIO SPECTRUM MANAGEMENT

The Impending Enforcement Problem

TOBY YOUELL

Contents

Several tens of thousands of people around the world work in spectrum management. As the capabilities of artificial intelligence (AI) continue to impress and improve, should they be worried about being made redundant? This chapter argues that the answer is no: radio spectrum management cannot be "outsourced" to software. In fact, despite its promising capabilities, AI poses difficult problems that might provide spectrum managers with many headaches (and projects) in the years to come.

Artificial Intelligence

As the brain appears to be a super-computer, computer experts have long wondered if a super-computer could be considered a type of brain. In 1950, Alan Turing wondered if an artificial machine might ever fool a human into thinking it was also a human, the so-called Imitation Game. Ever since the 1950s, researchers have experimented with such AI. The field has gone in and out of fashion several times in the last decade. In the 2010s, excitement grew around the technology for three reasons:

1. improvements to "machine learning" algorithms, such as the use of neural networks, that use ideas from neuro-science to develop ways to teach machines;
2. huge amounts of new data points being created that can be used for machine learning training; and
3. computing power from super-computers that can be accessed over the internet, as well as mass-market GPUs, that allow machine learning algorithms to be executed.

These three factors have enabled AI to surpass human intelligence in several discrete areas. In 2011, IBM's Watson beat previous human champions of the US TV quiz show Jeopardy.[1] In 2016, AlphaGo's neural network, whose machine learning training was not supervised by humans, beat the world champion of Go, the notoriously complicated Chinese strategy game.[2] In recent years, researchers have demonstrated the potential to deploy AI in a growing number of real-world situations that add real value.

In the developed world, AI can be considered ubiquitous, albeit invisible. Googlemaps uses it to optimise drivers' routes, Netflix uses it to recommend films to individual subscribers, and mass-market drones use it to keep their stability in the air. AI has proven so capable at such a broad range of challenges that it is increasingly relied upon for autonomous decision making when humans are not available. Because communications between Mars and Earth have an inevitable latency problem (it takes light several minutes to travel between the two planets), the US Space Agency, NASA, uses AI-powered Terrain Relative Navigation for its spacecraft to control their own descents onto the Red Planet.[3]

Radio Spectrum

Global spectrum management comprises a complex matrix of primary law, regulations, international treaties, best practices, and property rights. The spectrum cannot be used until these rules and rights are settled, and the assumptions that underpin them are subject to years of scrutiny and tend to be quite conservative. With AI, one could conceivably dispense of all this contention, delay, and airline point accumulation, and just rely on software that perceives its environment and can coordinate with other users of the spectrum.

In short, seeing as AI can outperform humans at almost anything, why not radio spectrum management?

In the early 2000s, there was a great deal of optimism that cognitive radio could make radio spectrum management more efficient. This vision did not replace property rights as the default method of spectrum management, but the promise of AI is breathing new life into such ideas.

The telecoms industry and AI are already intertwined. Much of the data used in AI are gathered by Internet of Things networks, while aspects of modern and future telecoms networks use AI predictive analytics capabilities to manage components in their networks. Machine Learning implementations of AI can also help radios better perform various functions, such as spectrum sensing, Radio Access Technology classification, signal classification, Medium Access Control protocol identification, Power Allocation, and Attack detection (jamming).[4] The utopian vision is to combine these capabilities with autonomous decision making to enable dynamic spectrum access.

This vision has already been implemented in the stratosphere. Before the project was abandoned in early 2021, Loon would provide mesh connectivity between its High Altitude Platform Station balloons and a Temporospatial Software Defined Network (TS-SDN). The TS-SDN incorporated environmental data, signal propagation models, per-node radio hardware/antenna configurations, spectrum regulations, and licensing information.[5]

The SDN relied on predictive software to calculate the future state of every node in the constellation to route traffic and interference coordination. Such automated fleet management enabled storm and obstruction avoidance, navigation in restricted airspace, and balloon stability. It was, according to Google, scalable and could even include satellites in its network.

Using such techniques on Earth might help overcome a central paradox of radio spectrum management: it can be incredibly scarce, as evidenced by the high prices it attracts at auctions, yet at the same time it can be unused most of the time.

Down to Earth

The US National Science Foundation (NSF), which distributes millions of dollars for research into computing, is one of the institutions interested in the promise of AI. It is pumping money into projects that enable devices that want to communicate to work out for themselves which frequencies to use and at what power levels. Its leadership believes that self-aware devices could enable spectrum access to become as efficient as internet routing. General internet traffic does not travel over wires reserved for particular users, so why should particular users have exclusive access to a given frequency, it is argued.

At the time of writing, the NSF is setting up a Spectrum Innovation Initiative that aims to promote "dynamic and agile spectrum use". This includes establishing a $25 million National Center for Wireless Spectrum Research for spectrum innovation and workforce development, national radio dynamic zones (for pilots), spectrum research activities, and education and workforce development.[6]

Prior to the Spectrum Innovation Initiative, another branch of the US government, the Defense Advanced Research Projects Agency (DARPA), was already funding research into this area. In 2019, it concluded its $2 million Spectrum Collaboration Challenge. This challenge pitted teams of AI engines against each other to successfully access spectrum to complete a number of tasks (such as data transmission). One of the challenges featured an incumbent that would appear periodically. If it received interference then points would be deducted from competitors. The AI engines quickly learned how to avoid the incumbent. These competitions took place virtually in the largest spectrum emulator ever built, the Spectrum Colosseum. Despite being virtual, the final was displayed publicly to great fanfare at an industry conference in Los Angeles. A key aspect of its brief was that spectrum access should be collaborative. To be as collaborative as possible, the teams agreed to communicate with each other using a newly invented Collaborative intelligent Radio Network Interaction Language. This challenge demonstrated that AI engines could communicate with each other at similar or better levels of efficiency than would have been possible using static spectrum assignments. They also demonstrated that they could quickly adapt to change, at least in the emulator.[7]

One of the finalists, Zylinium, is piloting its AI engine in the real world in a test-bed in Utah. It is using an FCC Innovation Licence. These licences have also been issued for researchers in New York City and North Carolina. It affords the licence holder free access to a swath of spectrum, including the same spectrum that operators spent billions of dollars obtaining in recent auctions. The Colloseum has been reconstructed at Northeastern University and is available over the internet as an "AI Playground" for researchers.

The US government is continuing this work studying Radio Frequency Machine Learning Systems, with a focus on how the technology could be useful for the US military.

So What?

AI promises to provide dynamic spectrum assignments that respond to real demand for radio spectrum at any given moment. It raises the possibility of using AI not only to use spectrum more efficiently but also to prioritise certain favoured communications, such as those related to safety of life. In doing so, it challenges some basic assumptions that spectrum managers have about creating rules in advance and even the idea of dividing up spectrum by service and/or user. It promises to achieve this without extracting billions of dollars from the industry in the form of auction prices and spectrum usage fees.

For decades, software defined radio (SDR) has inspired spectrum managers but its real-world traction has often disappointed. But AI refreshes this promise by offering a superior understanding of the radio environment and enhanced collaboration with other radios. It is conceivable that these capabilities can reduce cognitive radios' current extensive use of databases. AI-powered SDR, alongside ever cheaper hardware, raises the possibility that any user could access any frequency at any time and without the audit trail that comes with database usage. But while this might remove the problem of spectrum scarcity, could it create other hidden costs for society?

After all, despite the appearance of rather vociferous disagreements in international fora, radio spectrum management is a relatively collegial industry. Its actors generally trust each other to follow and to understand the rules, most of the time. But as technical barriers to spectrum illegality collapse, novel problems in spectrum management arise.

By way of analogy, it was relatively easy for large film studios to control access to their content when it was only accessible in cinemas. But when technology allowed media content to be stored and copied on CDs and DVDs, respect for intellectual property became a choice. In Western countries with large, formal, and legally compliant economies, shops would only sell the official CDs and DVDs. But in countries with large informal economies, it was easy to buy the pirated content off the street.

The question then relates to the capacity for each state to enforce their rules for spectrum. Do regulators need to prepare to deal with Napsters of radio spectrum access, routinely flouting the law and doing damage to the industry? With much of the world dependent on reliable spectrum access, be it for Global Navigation Satellite Systems, or police radios, the prospects of spectrum chaos could be quite troubling. The current institutions of spectrum management appear mostly unprepared to deal with such enforcement challenges.

For all its imperfections, the global framework for radio spectrum management is so effective that most of the people are unaware of its existence. With the advent of AI, it is possible that radio spectrum will become one of those dysfunctional areas of public policy that everyone has an opinion about, like the availability of affordable housing, or the expansion of airports.

Solutions

When technology is an enabler rather than a barrier to illegal behaviour, law enforcement becomes much more important.

For some spectrum managers, the solutions to these issues are likely to be the avoidance of change at all costs. But as the "Luddite" destroyers of textile equipment discovered, it is difficult to resist technical change. After all, AI's potential benefits in spectrum management are enormous and it would be a shame not to take advantage of them.

Fortunately, AI technology itself offers some solutions to these enforcement issues. For example, AI offers improved classification of interfering signals. This may make it easier for individuals to identify the offending device and to ask them to change their transmissions. Policymakers may consider rules requiring that devices transmit information about their own identities, or that devices maintain logs of their activities that can be checked after an interference event.

Until such measures have been demonstrated to be workable, regulators might be inclined to limit the authorisation of AI-enabled spectrum management to particular locations and in particular frequencies. Access to spectrum in the real world will, of course, enable AI to improve as it generates more data points with which to train itself.

In conclusion, the transition towards any AI-based spectrum management regime is likely to be long, contentious, and complex. In short, it's spectrum management as usual.

Notes

1. J. Markoff, February 2011. Computer Wins on "Jeopardy!" Trivial, It's Not'. *The New York Times*. www.nytimes.com/2011/02/17/science/17jeopardy-watson.html.
2. D. Silver, J. Schrittwieser, K. Simonyan, et al., 2017. Mastering the Game of Go Without Human Knowledge. *Nature*, 550, pp. 354–359. https://doi.org/10.1038/nature24270.
3. T. Weiss, February 2021. NASA's Perserverence Rover Lands on Mars: Here's How It Is Using AI. *Enterprise AI*. www.enterpriseai.news/2021/02/19/perseverance-rover-lands-on-mars-heres-how-it-will-use-ai/.
4. F. Shah Mohammadi, March 2020. *Artificial Intelligence in Radio Frequencies*. IEEE Signal Processing Society. https://signalprocessingsociety.org/publications-resources/blog/artificial-intelligence-radio-frequencies.
5. D. Northfield, February 2021. As Loon's Balloons Wind Down, What are the Prospects for HAPS Business Models? *Policy Tracker*. www.policytracker.com/spectrum-dashboard/as-loons-balloons-wind-down-what-are-the-prospects-for-haps-business-models/.

6. D. Standeford, February 2021. US to Launch "Spectrum Innovation Institute". *Policy Tracker.* www.policytracker.com/us-to-launch-spectrum-innovation-institute/.
7. T. Youell, November 2019. US Researchers Win $2 m in Spectrum Access Technology Competition. *Policy Tracker.* www.policytracker.com/us-researchers-win-2m-in-spectrum-access-technology-competition/.

10

WHY CAMPAIGNS AGAINST 5G HAVE BEEN SO SUCCESSFUL AND WHAT CAN BE DONE TO IMPROVE INDUSTRY MESSAGING

MARY LONGHURST[1] AND MARTIN SIMS[2]

Contents

For decades, individuals and organised groups have lobbied against mobile phones claiming they pose health risks, but the recent anti-5G campaigns have had an unprecedented impact. The penetration of these messages has been estimated at 50%[3] and never before have these campaigns led to arson attacks on base stations. In this chapter, we analyse the messaging on mobiles and health presented by both the anti-5G campaigners and industry. We compare these to academic research on running successful campaigns and make suggestions on how industry messaging could be improved.

In September 2020, as the consequences of the global coronavirus pandemic became increasingly apparent, the World Health Organization (WHO) and other UN bodies acknowledged that this the first pandemic in which technology and social media are being used on a massive scale to keep people safe, informed, productive and connected.[4]

DOI: 10.1201/9781003156765-11   **165**

But these UN bodies also acknowledged that information and communications technology (ICT) has enabled and amplified an infodemic that jeopardises measures to control the pandemic and enabled the proliferation of conspiracy theories. Indeed, the Covid-19 pandemic has led to a "parade of false, unproven and misleading claims about the virus making the rounds on social media, including allegations that 5G wireless technology somehow plays a role in the spread of the COVID-19 virus."[5] This has led to a spate of arson attacks on cell phone towers around the world, with the UK experiencing more than 70 incidents in May 2020 in 40 cases people have been attacked, either physically or verbally.[6]

In October 2020, the number of arson attacks on mobile phone masts had reached 87 in the UK, 50 in France, 30 in the Netherlands and 17 in New Zealand, according to the mobile industry body, the GSMA.[7] To put this into context, there are 66 million people who live in the UK, 67 million in France, 17 million in the Netherlands and 5 million in New Zealand.

Conspiracy theories are not a new phenomenon, and conspiracy theories in relation to mobile technology are not new either. But they have become more apparent as the ideas they propagate are spread more quickly and widely through social media.

The Power of Anti-5G Campaigning

Historically, campaigns about health risks associated with mobile phones used orthodox and lawful means but the arson attacks against 5G masts in several countries in 2020 marked a departure from accepted forms of non-violent protest. The anti-5G campaigns were so powerful that they inspired a small number of people to break the law and risk a custodial sentence.

Public opinion surveys showed the traction that the anti-5G message was achieving. A consumer poll in May 2020 found a fifth or more adults in 6 out of 14 countries surveyed agreed with the statement "I believe there are health risks associated with 5G."[8] This is despite advice from the WHO[9] and International Commission on Non-Ionizing Radiation Protection (ICNIRP)[10] that there is no conclusive evidence of adverse health effects from mobile phone usage if safety standards are followed.

The feeling of "uncertain times" created by the coronavirus epidemic, along with widely shared claims on social media that it was caused by 5G, certainly played a part in motivating the attacks. An April 2020 survey by UK regulator, Ofcom, found that 50% of respondents had seen false or misleading statements about 5G.[11]

However, in this chapter, we focus on the content of the anti-5G campaigns, seeking to establish the lessons that can be learnt for those that seek to reassure the public.

Analysing the Anti-5G Campaigns

Anti-5G messages and how they relate to emotional themes

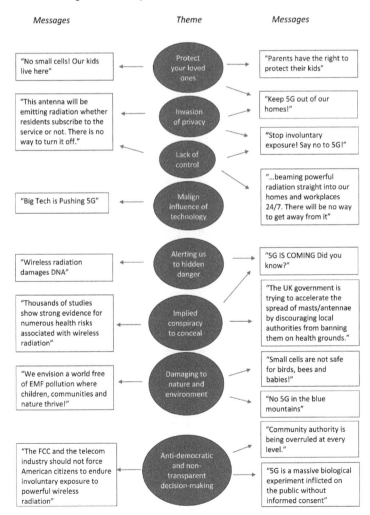

A key element of a successful campaign is its ability to connect with the target audience on an emotional level – does it stir "gut feelings"?

This was a prominent feature of anti-5G campaigns, which evoked strong emotional themes linked to powerful pre-existing narrative themes. We identified eight of these themes and explain their persuasive power as follows:

1. Protect your loved ones
2. Invasion of privacy
3. Lack of control
4. Malign influence of technology
5. Alerting us to hidden danger

The campaigns studied were produced by four anti-5G groups:

- Placards and flyers from Stop 5G International[12] – a global group that shares information and campaign materials with national campaigns
- Environmental Health Trust (EHT)[13] – a nonprofit organisation that lobbies extensively in the USA and takes legal action against the regulator, the FCC
- Say No to 5G UK,[14] a UK national campaign
- 5G crisis,[15] a US national campaign

The previous diagram gives examples of the messaging used by the anti-5G campaigners and how they evoked strong emotional themes.

Paradise Lost

The picture painted by the anti-5G messaging is of people who value nature and family living happily in local communities. But they are being attacked and need to protect their loved ones (*No small cells! Our kids live here*). The privacy which is key to their sense of security as individuals and families is threatened (*Keep 5G out of our homes!*). This is a malign and insidious force which they will not be able to control (*Stop involuntary exposure! Say no to 5G!*) and which is aligned with increasing technological automation and powerful business interests (*Big Tech is Pushing 5G*).

The campaign is trying to help and protect local communities by alerting us to this hidden danger (*5G IS COMING Did you know?*).

There is an implication that information has been concealed or with-held (*Thousands of studies show strong evidence for numerous health risks*) and that the natural justice we would expect from democratic insti-tutions is being by-passed (*Community authority is being overruled at every level*).

Furthermore, this is a serious issue not just for human beings but also for the natural environment with which we should have a harmo-nious relationship (*Honey bee says stop 5G*).

Linked together, these themes tell a powerful story, another aspect of successful campaigning.

Narrative Themes

The campaigners cast themselves as Robin Hood or Superman fig-ures, trying to save their communities from the "bad guys". Their por-trayal of 5G and the "Big Tech" forces behind it has echoes of the Superman enemy, Lex Luther, portrayed as both a mad scientist and owner of a huge corporation. The need to alert governments to hidden dangers perpetrated by powerful figures is also a familiar narrative theme, similar to Batman's relationship with the mayor of Gotham City. The need to live in harmony with nature and the apocalyptic consequences of failing to do so is a storyline in many films, such as *Waterworld* and *Interstellar*.

The narrative presented by anti-5G campaigners is very similar to what Jewett and Lawrence described as the "American Monomyth". This is an archetypal plot pattern in the US popular culture in which a community

> in a harmonious paradise is threatened by evil; normal institutions fail to contend with this threat; a selfless superhero emerges to renounce temptations and carry out the redemptive task; aided by fate, his deci-sive victory restores the community to its paradisiacal condition.[16]

It is notable that the anti-5G campaigns also present themselves as both the victims and heroes, a common theme of conspiracy theories:

> *Conspiracy theorists perceive and present themselves as the victim of orga-nized persecution. At the same time, they see themselves as brave antagonists*

taking on the villainous conspirators. Conspiratorial thinking involves a self-perception of simultaneously being a victim and a hero.[17]

Messaging Power

The success of the anti-5G activists meshes closely with academic research about effective campaigning. Campaigns are successful when they bring people together for a common purpose. A campaign proposition must be worthwhile to attract participants. This means there has to be clear benefits and solutions.[18] At a time when individuals and communities have a sense of losing control over their lives, anti-5G campaigning gives people a sense of agency and a sense of influence over the world around them. Like much political campaigning, the messages are designed to appeal more to the emotions than the intellect[19] and are primarily based on fear.

However, not everyone will decode anti-5G messages in the manner suggested earlier, depending on their own values and life experiences. To take the example of the UK, the surveys described earlier also show that while anti-5G campaigns achieved wide recognition, their persuasive power was limited. Those that thought 5G didn't have health risks were in fact the biggest group: 61% compared to 14% who thought there were risks.[20] With the UK study, as 50% of respondents[21] had seen anti-5G messages, we may conclude that 36% were not convinced.

Anti-5G campaigns seek to evoke pre-existing narratives with powerful emotional resonances, giving them a strong impact on some audiences but not all. For example, those who do not accept the anti-5G lobby's claims of health risks are likely to regard the emotional messaging as deceitful rather than persuasive.

Emotional impact and a strong narrative are key ingredients in winning someone over, and the anti-5G campaigns score highly on both.

They appeal to widely accepted values: family, selflessness, care for others, community, kindness, fairness, democracy, due process; values that emphasise morality and humanity. This is in sharp contrast with values often cast as representing the opposite: speed, efficiency and technology. This is typically the focus of industry messaging and to examine this we have taken the approach of the UK mobile operators as a case study.

5G Industry Messaging in the UK

The four main UK players are Vodafone, EE, O2 and 3. A review of their 5G campaign materials – websites, billboards, social media campaigns – reveal powerful promotional initiatives. The Vodafone campaign focuses on "It's more . . . speed, spontaneity, friends". This links to a #5GameChanger social media programme and the use of celebrities and sports personalities to promote the benefits of 5G through engaging videos.

EE's campaign focus is "We're 5GEE, Are you?". It features a series of videos from "ambassadors" designed to appeal to a younger audience.[22] The focus again is on the benefits of connectivity and speed. The O2 campaign welcomes you to the 5G generation heralding its ability to empower and offer more reliable connection. 3's campaign again focuses on speed and quality with a specific mention of the benefits to gamers.

The fact that all these campaigns focus on the benefits is not surprising. They are high investment marketing campaigns that understand their market and the value of engaging and entertaining through interactive campaigns that are emotionally influential and carefully focused on their target market.

What is surprising is the level of 5G safety communication in comparison. On the Vodafone, EE and O2 sites, the safety information is limited to a few paragraphs of text, linking to other organisations, such as the WHO, and can be hard to find on their websites. 3 is the exception when it is more prominent and more detailed. All the same, when considering that this is an issue that has led to cases of employees being attacked during a pandemic when maintaining connectivity within communities is so vital, it's surprising that the same marketing tools have not been applied to this issue.

A campaign by Vodafone in New Zealand shows the power of applying these marketing tools of segmentation and engagement to an education campaign is ran to tackle these 5G conspiracy theories. This is explained in more detail in the next chapter by Nicky Preston.

By working with influential members of the local community, and responding to the online conspiracy theories with factual information that was designed to be funny and shared, Vodafone New Zealand saw the arson attacks stop. This points to the potential for utilising the power and reach of social media channels, but developing content with a local community that resonates among them, to negate some of the online impacts in a world dominated by social media use.[23]

Theoretical Models

These 5G statements rely on an information strategy.[24] A one-way communication model aiming at relaying information through a company-centric process that involves "telling, not listening". It's a "decide, dictate, defence" approach[25] where the company decides on the action, dictates the terms, and then defends the action. It has long been understood, however, that in terms of social marketing, this approach no longer works.[26] Indeed, the telecommunications companies understand that just providing information will not build trust and sell their products and services, so why use it to influence the public on an issue of safety?

There is powerful evidence to suggest that messaging in this area would be favourably received by business.

Safety advice and statements have been issued by a number of health and industry bodies: the WHO, European Union, the ICNIRP and national regulators. These are all establishment figures against which conspiracy theories often rally. Indeed, the Edelman Trust Barometer for 2020[4][27] reveals that the public expects businesses to be more involved in societal issues, especially where others are failing. This level of trust is higher compared to government, media and NGOs with respect to ethical issues and competence. Furthermore, the Edelman research claims that business gains the most trust by being a guardian of information quality. There is clearly a significant role to be played, therefore, by the telecommunications companies.

This missed opportunity highlights the main difference in managing commercial marketing versus social marketing. Commercial marketing puts the corporation and its shareholders at its centre (Hastings, 2012), in the pursuit of profit. Social marketing, in comparison, puts the citizen at its centre. Both may take a multi-stakeholder approach (Freeman, 1984), appreciating the influence of a network of diverse, interested parties, but the central focus is different. This is a significant shift for a marketing team to make and, as we see, one of the major telecommunications providers in the UK, haven't so far made. If they had, they would have communicated more powerfully positive messages of cohesive communities and enhanced social ends, while strongly dismissing misinformation and messages of potential harm.

These communication gaps reveal the pitfalls of the traditional commercial marketing approach in dealing with social issues. They prove that, despite a multi-stakeholder approach, stakeholders do not have equal power over an organisation (Fleming and Jones, 2013) which impacts their ability to voice their interests and which can lead to the type of frustration and dis-trust apparent in the anti-5G activity. Small groups thus form to vent their feelings in a way that is highly visible and, in this case, potentially dangerous. By allowing marketing teams to prioritise their focus on the consumer as a means of achieving sales and profit, companies ignore the fact that those consumers are also citizens with social concerns. The same citizens telecommunications companies face when they want to erect a phone mast near their homes.

A company is more than just one issue and one new product or service. It's a complex network of activities and communications, both within the organisation and externally. This means ensuring its actions are perceived across all its touch points, by employees and the local community alike. Viewing the launch of new technology through the unilateral lens of consumer benefits is likely to become an increasing problem as communities become increasingly health and environmentally aware. This means viewing the marketing of a company and its products and services as a whole.

Increasingly, campaigning organisations and researchers are understanding from successful campaigns relating to climate change and other social issues, that changing behaviour demands the use of the same powerful marketing insights and creativity that are used to sell products and services. The use of carefully crafted, specific messaging, the use of influential messengers, and the engaging and entertaining use of interactive media channels is a powerful way to influence and change behaviour. This demands time and financial investment in campaigning, however. Yet, considering the increase in trust now held with business,[28] there is a significant opportunity awaiting companies which makes it an investment worth making.

Notes

1. Email: mlonghurst@epochcomms.com.
2. Primary contact author, email: martin@policytracker.com.
3. See Ofcom, 21 April 2020. *COVID-19 News and Information: Consumption and Attitudes.* Discussed in more detail below.
4. World Health Organisation, 2020. *Managing the COVID-19 Infodemic: Promoting Healthy Behaviours and Mitigating the Harm from Misinformation and Disinformation.* www.who.int/news/item/23-09-2020-mana ging-the-covid-19-infodemic-promoting-healthy-behaviours-and-mitigating-the-harm-from-misinformation-and-disinformation.
5. A. Gruzd and P. Mai, 2020. Going viral: How a Single Tweet Spawned a COVID-19 Conspiracy Theory on Twitter. *Big Data & Society*, 7(2).
6. I.A. Hamilton. 2020. 77 Cell Phone Towers Have Been Set on Fire Due to a Weird Coronavirus 5G Conspiracy Theory. *Business Insider.*
7. Figures given by a representative from the GSMA in a webinar entitled 'Effective EMF Communications' on at On 14 October 2020. See www. gsma.com/gsmaeurope/events/9th-gsma-europe-emf-forum/.
8. Deloite, 2021. *TMT Predictons*, p. 25, USA: Deloite.
9. See www.who.int/news-room/fact-sheets/detail/electromagnetic-fields-and-public-health-mobile-phones.
10. See www.icnirp.org/en/applications/mobile-phones/index.html.
11. Ofcom, *COVID-19 News and Information.*
12. See https://stop5ginternational.org/.
13. See https://ehtrust.org/about/.
14. See www.saynoto5g.uk/.
15. See www.5gcrisis.com/toolkit.
16. J.S. Lawrence and R. Jewett, 2002. *The Myth of the American Superhero*, p. 6, MI, USA: Wm. B. Eerdmans Publishing Company.
17. S. Lewandowsky and J. Cook, 2020. *The Conspiracy Theory Handbook.* http://sks.to/conspiracy p7.
18. C. Rose, 2010. *How to Win Campaigns.* Abingdon, Oxon: Earthscan.
19. T. Brader, 2006. *Campaign for Hearts and Minds: How Emotional Appeals in Political Ads Work.* Chicago: University of Chicago Press.
20. Deloite, TMT Predictons, p. 26. Those that strongly agreed, or tended to agree with the statement "I believe there are health risks associated with 5G," amounted to 61% in the UK, whereas those that strongly disagreed, or tended to disagree amounted to 14%.
21. Ofcom, *COVID-19 News and Information.*

22. Described as "'Lee the Sneakerhead' who uses faster speeds to collect limited edition trainers before anyone else. Sojourn the Skate Queen who teaches the nation to skate through seamless connections and The Ember Trio, the hip-hop cellists." See www.thedrum.com/creative-works/project/wunderman-thompson-uk-were-5ge-are-you.
23. N. Preston, 2020. 5G and Covid-19 Conspiracy Theories; How Vodafone New Zealand responded to cell tower arson attacks by using humour to beat rumour.
24. J.E. Grunig and T. Hunt, 1984. *Managing Public Relations*. New York: Holt, Rinehart and Winston.
25. T. Watson, S. Osborne-Brown, and M. Longhurst, 2002. Issues Negotiation – Investing in Stakeholders. *Corporate Communications*, 7(1), pp. 54–61.
26. Ibid.
27. Edelman Trust Barometer, 2020.
28. Ibid.

References

Fleming, P. and Jones, M. T. 2013. *The End of Corporate Social Responsibility*. London, Sage.

Freeman, R. E. 1984. *Strategic Stakeholder Management*. Minneapolis, University of Minnesota Press.

Hastings, G. 2012. *The Marketing Matrix*. London, Routledge.

11

5G and Covid-19 Conspiracy Theories

How Vodafone New Zealand Responded
to Cell Tower Arson Attacks by Using
Humour to Beat Rumour Online

NICKY PRESTON

Contents

Introduction

Social media channels provide a mechanism to connect globally and spread positive news; however, there is also a dark side that governments and corporations are facing with regards to online conspiracy theories having offline impacts. Vodafone New Zealand needed to confront this in 2020 when misinformation linking Covid-19 and 5G led to a spate of arson attacks on mobile phone masts, centred in a certain geographic location in South Auckland. A total of 88% of adult New Zealanders regularly use YouTube, while 84% use Facebook[1],

DOI: 10.1201/9781003156765-12

177

highlighting the scale of use – with Māori as tangata whenua (indigenous New Zealanders) particularly relying on these channels as a source of news.

This chapter will examine the issue of Covid-19 and 5G conspiracy theories online, with "'Totally false' 5G conspiracy theories" blamed for a series of arson attempts on 5G cell towers around New Zealand.[2] Secondly, it will explore how certain communities in New Zealand, particularly Māori and Pacific Islanders, have been found to be more susceptible to conspiracy theories via social media channels due to years of mistrust in the system and traditional media outlets. Thirdly, research will show the scale of the problem in relation to misinformation about Covid-19 spreading online – before explaining how Vodafone New Zealand addressed this misinformation, both alongside an industry association and on its own social media platforms, to try to 'fight fire with fire'.

Analysis of social media use shows that some communities, particularly Māori and Pacific Islanders as two of the minority groups in New Zealand and highly represented within South Auckland, are more reliant on Facebook and online channels for their news and, therefore, more susceptible to misinformation – as well as suffering more from the digital divide.[3] Looking to Taiwan, as a country that has been relatively successfully weathering the pandemic and associated misinformation, Vodafone New Zealand devised a campaign to adopt the concept of using 'humour over rumour' online. This focused on developing educational videos, co-creating online content alongside experts from the Māori and Pacific Island communities based in South Auckland, to address 5G and Covid-19 misinformation.

As well as providing some considerations around why online misinformation spreads, this chapter will highlight how increased education measures using social media can be effective in combating misinformation on 5G, particularly among at-risk communities.

Social Media Use in Aotearoa New Zealand

New Zealand has a relatively high internet penetration and social media use. Research from early 2020[4] shows the country has a population of 4.8 million with an urbanisation rate of 87%. The country has

4.47 million internet users, or a penetration rate of 93%, and a staggering 6.49 million mobile phone connections. The number of active social media users is 3.6 million, which equates to a 75% penetration rate but growing at 3.5% each year. Like many Western nations, the country's citizens spend a considerable amount of time on social media platforms, calculated to be 1 hour and 45 minutes on average daily. The average number of social media accounts per user is 7.2.

This research also indicates the top five most-used social media platforms in New Zealand, in order noting the percentage of social media users who said they used each platform in the past month, are YouTube (88%), Facebook (84%), Facebook Messenger (72%), Instagram (52%) and Pinterest (35%).

The Issue of Misinformation on Social Media Channels

In 2016, Oxford Dictionaries named 'post-truth' as the word of the year, defined as the adjective "relating to or denoting circumstances in which objective facts are less influential in shaping public opinion than appeals to emotion and personal belief", after its use spiked 2,000% compared to the prior year.[5] Post-truth, or misinformation, reduces the effectiveness of social media's information content, often making these platforms unreliable as news sources.[6]

Social media inherently includes the ability to disseminate user-generated content, which is often not subject to fact-checking as traditional media would have been.[7] Due to the networked nature of platforms, with users sharing information with people similar in views to them, an "echo chamber" can be created whereby selective exposure to content generates homogeneous clusters of people who share and perpetuate misinformation online causing false truths to further spread.[8] Furthermore, researchers[9] have highlighted a lack of user understanding of privacy concerns mean that users often don't realise their interactions online can be shared much wider than intended.[10]

As the coronavirus crisis started to spread across the globe in early 2020, the World Health Organization (WHO) warned[11] "we're not just fighting an epidemic; we're fighting an infodemic". It cautioned that "fake news spreads faster and more easily than the virus, and it is just as dangerous". John Cook, an expert on misinformation with

George Mason University's Center for Climate Change Communication, also highlighted[12] that a "calamitous event" like the pandemic creates a "very fertile breeding ground for conspiracy theories".

The challenge for governments and organisations is how to respond. A Yale University workshop in 2017 on misinformation suggested that critical thinking training in schools would be more effective than regulatory actions like fines for misbehaviour.[13] In 2018, the European Commission set up a high-level group of experts, the High-Level Expert Group on Artificial Intelligence,[14] to provide guidance on policy initiatives to counter fake news and misinformation online. One key recommendation[15] was to ensure the privacy-compliant sharing of data. However, when in 2020 Facebook called for international regulation of content, European lawmakers openly rejected this instead suggesting platform-implemented standards and systems.[16]

Realistically, regulatory measures can be challenging for large governments, let alone small countries like New Zealand with its relatively tiny population. The differing responses by Google and Facebook to the proposed Australian news code around paying for content[17] emphasise the high stakes for tech giants and the variation in platforms' willingness to respond. New Zealand lawmakers will be keenly watching these events unfold, particularly because local policy positions often follow an Australian response. And despite positive efforts to engage various platforms around eliminating extremist content online following the Christchurch terror attacks, including a global call to action attracting 48 countries,[18] regulatory changes in New Zealand are still yet to pass.

Online Misinformation around Covid-19 and 5G Causing Real-Life Impacts

There was a clash between online and offline impacts in many countries when Covid-19 conspiracy theories online led to a spate of arson attacks on cell phone towers around the world.[19] In May 2020, telecommunications industry group, GSMA,[20] reported ten European countries had experienced arson attacks on infrastructure such as mobile phone masts, or seen assaults on maintenance workers.

In New Zealand, these arson attacks began shortly after the country went into nationwide lockdown on March 25, 2020. Interestingly, the vast majority (15 attacks) occurred in or near South Auckland[21] as one of the districts of New Zealand's largest city, an area with historic issues of disadvantage. The arsons were directly linked to Covid-19 and 5G conspiracy theories when one of the first attacks was filmed and the video posted to Facebook,[22] with the arsonist referencing the pandemic on camera. While Facebook took down the clip once alerted to it, the *NZ Herald* outlet highlighted that "before the account was blocked, the clip was shared hundreds of times – and members of the local telco industry are on high alert that similar content could spur similar attacks". A month later, following more arson attacks on cell phone masts in South Auckland, another local newspaper[23] quoted analyst Dr Paul G. Buchanan saying that the "megaphone of social media was enabling conspiracy theories to quickly gain apparent credibility and manifest in physical acts such as the anti-5G attacks".

The level of concern around this Covid-19 misinformation and cell phone mast attacks led to the New Zealand Prime Minister, Jacinda Ardern, being questioned in early April by reporters on whether the government believed there were any links between Covid-19 and 5G, to which she quickly dismissed as false.[24] "That is not true," she said. "I can't state it clearly enough. I almost hesitate to speak to it on this platform – it is just not true". However, the attacks continued – and two months later, when asked by the *NZ Herald*, Vodafone New Zealand infrastructure director Tony Baird explained the company had "been introducing additional security measures such as surveillance cameras in a bid to protect cell sites".

While telecommunications companies like Vodafone New Zealand were beefing up security measures, there was also an apparent need to address the online misinformation. The platform owner, Facebook, had come under local scrutiny and, when asked what the platform was doing to try to combat this misinformation, Facebook asserted[25] that "We're taking aggressive steps to stop misinformation and harmful content . . . we have also begun removing false claims that 5G technology causes the symptoms of or contraction of Covid-19".

However, additional concerns were soon raised about Covid-19 conspiracy theories circulating online and the seemingly related protest rallies that were occurring around the world and in New Zealand. As reporter Marc Daalder noted in the online news site Newsroom.co.nz,[26] "Facebook's inability to rein in misinformation has led to demonstrable harm and the exacerbation of the Covid-19 pandemic". The article went on to quote Dr Catherine Strong, a senior journalism lecturer at Massey University who studies social media and fake news, who said:

> People who would not have normally gone to seek that type of information are getting it. People who would never protest are protesting against lockdown, saying it's against their rights, because things on Facebook have given them the emotional feeling that this is all made up anyway and that it's been overstepped.

When looking at Covid-19 and 5G misinformation, a worst-case outcome would be an arson attack on a mobile phone mast caused significant damage and resulted in a loss of connectivity for the community in which it is situated – which the New Zealand industry association the Telecommunications Forum warned about as the arson attacks continued.[27] However, the best outcome was preventing these sorts of attacks – with the hope being that if the misinformation that is causing the attacks to occur was addressed, then there would be both offline and online benefits.

Misinformation as a Greater Problem among Māori

Within New Zealand, misinformation on social media platforms is recognised to be more of a challenge among Māori, leading to the need for business and policy responses to include a specific focus on education for indigenous New Zealanders.

As noted earlier, the mobile phone mast arson attacks predominately took place in South Auckland – with 15 of the eventual total 18 attacks in 2020 taking place in this area – where Māori and Pacific Islanders are represented at a much higher rate. A locality population snapshot[28] from 2014 shows that 19% (52,236) of the South Auckland Whānau Direct area population identified with Māori ethnicity, 33% (90,717) with Pacific ethnic groups, 18% (49,893) with Asian ethnic groups, 20% (55,233) with European ethnic groups and 10% (23,169) with Other ethnic groups.

Commentators have highlighted there is a historic lack of trust among Māori in traditional news sources,[29] which has led to conspiracy theories around 5G and Covid-19 flourishing, as there is a greater reliance on social media channels as a form of news – both as an information source and via virtual word-of-mouth. Research shows Māori often view Facebook as a quasi-newspaper and place to learn about general news[30] – partly because Māori place greater emphasis on storytelling, with social media channels replicating the form. The context of why racial disparities exist is important, as it's acknowledged that Māori can display a mistrust of research, originating from a historical experience of exploitation and violation.[31] This dates back to when New Zealand was first formed as a nation, and misinformation against Māori has been reportedly perpetuated since in the 1860s over land disputes in the Waikato.[32]

When looking at device use, research conducted by the New Zealand Ministry of Māori Development[33] showed Māori (particularly young Māori) are over-represented amongst New Zealanders who are heavy and extensive users of new and emerging devices such as cell phones. This shows 85% of Māori aged 15–24 years use the internet for social media networking – with indigenous data sovereignty, or the right of indigenous people to govern the ownership and application of their own data harder to control in an online world. An issue that is even more pronounced among children.

A recent Netsafe report[34] explored the online attitudes and internet use of New Zealand children aged between 9 and 18 years. The representative sample included a cross section of ethnicity, including NZ European/Pākehā (73%), Māori (26%), Pacific (13%), Asian (15%) and "other ethnicity" (2%). The statistics highlighted a number of issues for Māori children in their access and response to the internet. Interestingly, Māori were more knowledgeable about managing privacy settings than other ethnicities, with Māori (66%), Asian (64%), Pākehā (63%) and Pacific (59%) being confident about this specific skill. However, when it came to fact checking information found online, Māori and Pacific Islanders were more vulnerable. When asked whether they find it easy to check if the information they find online is true, Asian (60%) and Pākehā (58%), Māori (56%) and Pacific (52%) indicated that this was fairly or very true.

Further governmental research indicates discriminatory policies disadvantage certain groups in Aotearoa New Zealand. The first Māori Social Survey, Te Kupenga 2013, shows while racial discrimination was the most commonly reported form, Māori also experience discrimination on the basis of other grounds, including age, gender and income.[35] This was emphasised in late 2020 when the major New Zealand news outlet Stuff.co.nz released[36] "a major internal investigation [uncovering long-term] evidence of racism and marginalisation against Māori". While the Human Right Act 1993[37] deems this unlawful, the reality is that large disparities exist with regards to the experiences of Māori in New Zealand. In this context, it can be expected that discrimination also exists in the online world – and the issues of misinformation are heightened with the lens of institutional distrust.

Other Education Campaigns to Engage the Māori Community

One challenge of targeting Māori, and related Pacific Island communities, with education via traditional news sources is this may perpetuate the historic lack of trust. If distrust is inherent in the Māori view of the world, then successful campaigns highlight that education materials and content should addressed from within the community, instead of being imposed on the group.

A review of an effective education campaign from NZ Fire and Emergency[38] to reduce unwanted fires highlighted that a 'native theory' approach in community development can promote a focus on the community's strengths and cultural processes, and advocates that projects are tailored to the specific community for maximum effectiveness. Another successful education campaign to engage Māori included efforts to control tobacco use. Education materials were created, and a policy was developed that utilised the role of indigenous leadership at a tribal, political and national advocacy level, which was found to be critical to shifting attitudes and resulting in a reduction of smoking among Māori.[39]

Humour is also documented to be received positively in advertising campaigns targeting Māori,[40] with 2015 research indicating humour "was understood as a way of selling the message". However, participants objected to the lack of consultation with Māori about the production of the advertisements, showing the importance of engaging the community which is being communicated with to ensure the effectiveness of the message.

When looking afield at successful online campaigns to combat misinformation online and incorporate humour, Taiwan provided a good example, using memes to fight Covid-19 pandemic rumours. In an interview following a number of proactive campaigns, Digital Minster Audrey Tang explained[41] one particularly effective approach. When local netizens started circulating a rumour that Taiwan would soon run out of toilet paper, within hours the Digital Ministry had released an infographic featuring Taiwan's Premier Su Tseng-chang debunking this rumour via a humorous caricature of him shaking his backside, playing on the similar-sounding Mandarin words for stockpile and bum. A key element was understanding the local humour and developing viral content that would likely be appealing to the target community – even if this meme may be confusing to non-Taiwanese nationals.

Taking inspiration from this, and taking into account that the conspiracy theories about 5G and Covid-19 had led to a spike of arson attacks in South Auckland that the company wanted to stop, Vodafone New Zealand devised a social media education campaign alongside local community members to engage the target audience, have viral appeal and aim to debunk myths. While this was aimed specifically at Māori and Pacific Islanders living within the geographical area, misinformation effects all New Zealanders. By addressing this issue among the most effected community, it was hoped that the broader appeal of the videos would create more positive impacts in terms of educating and protecting more online New Zealanders.

Case study: Vodafone NZ Uses Humour to Beat 5G Rumours among South Aucklanders

Vodafone was the first telecommunications company to launch a commercial 5G mobile network in New Zealand[42] – with the next generation network officially switched on in December 2019. Vodafone New Zealand was also one of three local mobile operators who experienced arson attacks on its digital infrastructure in South Auckland, linked to Covid-19 and 5G conspiracy theories. Following a high number of attacks over a short time period, the company was quoted in a joint statement[43] issued by the NZ Telecommunications Forum denouncing the attacks. Here Infrastructure Director Tony Baird said:

> These attacks are infuriating and can have real connectivity impacts for New Zealanders – meaning people could have reduced mobile phone and internet coverage in an area with a damaged cell site, which is a real issue particularly in South Auckland. While we've been able to keep customers connected so far, each attack has a cumulative negative impact.

To strengthen the local industry response to the misinformation, Vodafone New Zealand advocated for and helped contribute to a factual website (www.5gfacts.org.nz) launched by the NZ Telecommunications Forum in September 2020. This website aims to collate scientific information about 5G to provide a 'one stop shop' of reputable sources for any concerned citizens. This 5G Facts website was developed following research[44] that showed only a quarter of New Zealanders feel they know a reasonable amount about 5G mobile technology, and 86% said would like easier ways to learn more factual information. The website included links to fact sheets produced by the Prime Minister's Chief Science Advisor, the New Zealand Ministry of Health, the Ministry of Business, Innovation and Employment (MBIE) and the WHO. As Telecommunications Forum Chief Executive, Geoff Thorn noted when the site launched:

> There are lots of claims made about 5G, especially on the Internet and social media, and it's often hard to separate fact from fiction. If you want to learn more about 5G, we strongly encourage you to use respected sources of information from science and health experts. So, we've made www.5gfacts.org.nz as a 'one-stop shop' where you can get basic information and easily link to reputable expert sources.

At the same time, Vodafone New Zealand recognised there was a need to take this one-step further and devise more ways to share the factual website as widely as possible on social media to the target audience – as the place where rumours were circulating. The company

contacted South Auckland-based community and content development agency, Ngahere Communities, in June 2020, at the height of
the arson attacks, to engage them to help fight the spread of misinformation among Māori and Pacific Island communities online.
Engaging a locally based organisation was incredibly important to the
project, as part of Vodafone New Zealand's commitment to working with indigenous groups and a policy it launched in July 2020 to
'Honour the principles of the Treaty of Waitangi',[45] as New Zealand's
founding document.[46]

The first stage of the project was research, and Ngahere Communities set up a series of wananga (workshops) to determine what misinformation existed in the South Auckland community around 5G and
Covid-19, with a view to eventually creating content to debunk these
myths. Three in-person workshops were held, each lasting approximately one hour with a cross section of demographics – including
technology professionals, local community champions, family representatives and high school children (as heavy users of social media).
The workshops were originally planned for late-July but were eventually delayed until September after Auckland needed to go into a
regional lockdown when Covid-19 cases re-emerged. These workshops
involved Vodafone New Zealand first sharing basic info on what 5G
is, and isn't, and how it's being rolled out. Then Ngahere Communities led korero (conversations) on the opportunities 5G might bring to
South Auckland communities; the concerns the communities might
have about 5G; what information communities need about 5G, and
where this information should be shared.

Some of the key themes that came out of the discussions included
(Table 11.1):

Table 11.1   Key Themes from Workshop Discussions

OPPORTUNITIES	CONCERNS	QUESTIONS
– New and improved ways to stay connected with whanau (family).	– There is a lack of under-standing about 5G and lots of misinformation circulat-ing, particularly through social media.	– What has New Zealand learnt from other countries that have rolled out 5G?

(Continued)

Table 11.1   (Continued)

OPPORTUNITIES	CONCERNS	QUESTIONS
– Increased opportunity for online learning, education and skills development.	– Access to 5G and the devices it enables is mostly closed off to these communities (due to unaffordability), the digital divide is about to get wider.	– Will there be a forced upgrade path?
– Opportunity to evolve traditional activities online.	– There are low levels of trust, particularly when it comes to health, safety, security and access.	– Who owns 5G and can access be free or cheap?
– FUN! Faster means better for gaming and recreational activities. – Efficiency for business and working from home. – Creates new jobs and opportunities for skills development. – Greater opportunities for UX, big data, human interaction, robotics, AR/VR, medical and information sharing.	– How and where we educate the community is important.	– How will this increase younger generations' addictions to technology?

Taking these insights, and the concept from Taiwan of using humour to beat rumours on social media, Vodafone NZ commissioned Ngahere Communities to create a three-part series of online videos. Online influencer and local South Auckland personality, Torrell Tafa, was engaged to feature in these videos and share the content via his social media channels first, to authentically present 5G concepts to the community – by the messages coming from someone who had already built trust within the community. To foster more confidence among an audience that lacked trust in traditional sources, a Māori industry tech expert, Nikora Ngaropo, a Member of the Digital Council for Aotearoa New Zealand, was engaged to communicate factual messages about the safety of fifth generation mobile network technology (5G).

The key focus was on trying to address an information gap and help New Zealanders, particularly South Aucklanders, to better understand the potential of 5G. New Zealanders increasingly rely on digital infrastructure to keep us connected to the internet, but it was acknowledged that the telecommunications industry tends to over-complicate how mobile network technologies such as 5G operate. The science can be technical and confusing, so these videos attempted to translate complex topics on how phone and internet networks work into easily digestible videos and shared on social media – explaining electromagnetic frequencies and the evolution of mobile tech.

When launching the videos, Kirstin Te Wao, Head of Māori Development at Vodafone New Zealand noted,

> Part of the challenge the tech sector has is a lack of cultural diversity within key decision making roles that can translate into messaging that doesn't resonate with many of the people within our communities. Partnering with an organisation like Ngahere Communities means that we bring community concerns, voices and solutions to the fore in a meaningful way that resonates with them. As part of our commitment to Te Tiriti o Waitangi, we're working hard to create meaningful, enduring and authentic relationships with Māori innovators and entrepreneurs like Manawa and her team.

Manawa Udy, CEO of Ngahere Communities, who led the campaign alongside her team including producer Jason Manumu'a, added,

> We partnered with Vodafone on this project because we agreed with the kaupapa around more inclusive and accessible 5G education, and the concept of including our community in these bigger conversations. Too often our communities are expected to just 'get on board' with new ideas and opportunities but we need to journey through these big changes together, it helps to reduce the amount of fake news and give facts, not fiction. It's no secret that South Auckland is dripping with creative talent, and we were stoked to have an opportunity to get our crew together, including creator and social media influencer Torrell Tafa, to work on this project. The result is three choice videos that we hope are both engaging and informative.

Nikora Ngaropo, Member of the Digital Council for Aotearoa New Zealand, explained his involvement and stated,

> I'm glad I've been able to bring a technology voice to this piece, as a member of the Digital Council for Aotearoa New Zealand. I think it's

important that our communities understand facts and don't get caught up in misinformation which is so easy to do these days. Also Ngahere Communities was great to work with, it's important that we have businesses involved that are connected to the pulse of our communities and know how to engage them.

The three videos were shared online by content creator Torrell Tafa on his Facebook, Instagram and Tiktok channels and then on the Vodafone New Zealand social media pages. Each was approximately 2 minutes long and, via captions, encouraged people to head to www.5gfacts.org.nz to find out more about how 5G really works and included:

1. That 5G's so extra, a video that was shared on, Friday, 20 November. The first video focused on explaining the potential of 5G and the evolution of mobile networks, from the first generation to the fifth generation, via a running race to show how speeds had evolved over time.

2. 5G Myths, the first of two myth-busting videos addressing some of the more prevalent online conspiracy theories, was first shared on, Friday, 27 November. This humorous video focused on communicating how electromagnetic frequencies work.

3. 5G versus Nature shared on, Thursday, 3 December. This explored some of the more frequent myths that the community had raised, linking 5G to fears about the environment including the health of birds and plants.

The engagement on each video was strong across both Torrella Tafa's social media pages and the Vodafone New Zealand channels. Most importantly, no arson attacks have been recorded since the videos were posted online.

Conclusion: Addressing Misinformation on Online Channels among a Certain Community

Social media platforms rely on a user-generated content, which isn't subject to fact checking like traditional forms of media with regulatory oversight, and where necessary corrections are printed if an incorrect fact is uncovered. The global nature of social media platforms and New Zealand's relative small size makes regulatory measures challenging, and differing platform owners' response to Australia's proposed changes on news content highlights the difficulty a single country faces to enforce major, often costly, changes on a corporation with billions of users and a global reach. Therefore, tapping into the power of social media campaigns via local education measures can be a successful option when responding to misinformation online.

As established, indigenous New Zealanders Māori are more at risk from misinformation on social media as they not only lack trust in traditional media sources but also display higher use than other online New Zealanders. The potential for negative offline impacts from online misinformation was evident when Vodafone New Zealand and the other mobile operators in the country experienced a spate of arson attacks on mobile phone masts when misinformation around 5G and Covid-19 started spreading online, particularly impacting the South Auckland community with a high Maori population.

Education campaigns can be challenging to track the effectiveness in terms of social media metrics. But by working with influential members of the local community, and responding to the online conspiracy theories with factual information that was designed to be funny and shared, Vodafone New Zealand saw the arson attacks stop. This points to the potential for utilising the power and reach of social media channels, by developing content with a local community that resonates among them, to negate some of the online impacts in a world dominated by social media use.

Notes

1. Digital 2020 Global Digital Overview, 2020. *Hootsuite and We Are Social.* www.slideshare.net/DataReportal/digital-2020-global-digital-overview-january-2020-v01-226017535.
2. TVNZ, 16 May 2020. www.tvnz.co.nz/one-news/new-zealand/totally-false-5g-conspiracy-theories-blamed-series-infuriating-cell-tower-arsons-across-nz.
3. J. Greenbrook-Held, 2011. *The Domestic Divide: Access to the Internet in New Zealand.* www.academia.edu/3731220/The_domestic_divide_Access_to_the_Internet_in_New_Zealand.
4. Digital 2020 Global Digital Overview, *Hootsuite and We Are Social.*
5. Guardian, 15 November 2016. www.theguardian.com/books/2016/nov/15/post-truth-named-word-of-the-year-by-oxford-dictionaries.
6. C. Budak, D. Agrawal, and A.E. Abbadi, 2011. *Limiting the Spread of Misinformation in Social Networks.* www.ra.ethz.ch/cdstore/www2011/proceedings/p665.pdf.

7. N.A. Karlova and K.E. Fisher, 2013. A Social Diffusion Model of Misinformation and Disinformation for Understanding Human Information Behaviour. *Information Research*, 18(1). http://informationr.net/ir/18-1/paper573.html.

8. M.D. Vicario, A. Bessi, F. Zollo, F. Petroni, A. Scala, G. Caldarelli, H.E. Stanley, and W. Quattrociocchi, 2016. *Echo Chambers in the Age of Misinformation*. www.pnas.org/content/113/3/554/.

9. S. Hinton and L. Hjorth, 2013. *Understanding Social Media*. http://sk.sagepub.com.ezproxy.auckland.ac.nz/books/understanding-social-media.

10. A.B. Albarran, 2013. *The Social Media Industries*. Routledge. ProQuest Ebook Central. http://ebookcentral.proquest.com/lib/auckland/detail.action?docID=1143700.

11. World Health Organisation, 15 February 2020. www.who.int/dg/speeches/detail/munich-security-conference.

12. Washington Post, 1 May 2020. www.washingtonpost.com/technology/2020/05/01/5g-conspiracy-theory-coronavirus-misinformation/.

13. The Information Society Project, 7 March 2017. *Fighting Fake News Workshop Report*. The Floyd Abrams Institute for Freedom of Expression. https://law.yale.edu/sites/default/files/area/center/isp/documents/fighting_fake_news_-_workshop_report.pdf.

14. European Commission, 2019. *Shaping Europe's Digital Future*. https://ec.europa.eu/digital-single-market/en/high-level-expert-group-artificial-intelligence.

15. European Commission, 12 March 2018. *Final Report of the High Level Expert Group on Fake News and Online Disinformation*. https://ec.europa.eu/digital-single-market/en/news/final-report-high-level-expert-group-fake-news-and-online-disinformation.

16. C. Zakrzewski, 28 February 2020. The Technology 202: Mark Zuckerberg's Icy Reception in E.U. Signals Coming Clashes Over Tech Regulation. *Washington Post*. www.washingtonpost.com/news/powerpost/paloma/the-technology-202/2020/02/18/the-technology-202-mark-zuckerberg-s-icy-reception-in-e-u-signals-coming-clashes-over-tech-regulation/5e4ae6e4602ff12f6a67114c/.

17. BBC, 19 February 2021. *Australia News Code: What's This Row with Facebook and Google All About?* www.bbc.com/news/world-australia-56107028.

18. www.beehive.govt.nz/release/significant-progress-made-eliminating-terrorist-content-online.

19. A. Satariano and D. Alba, 10 April 2020. Burning Cell Towers, Out of Baseless Fear They Spread the Virus. *NY Times*. www.nytimes.com/2020/04/10/technology/coronavirus-5g-uk.html.

20. GSMA Europe, 19 May 2020. *Joint Statement of the UNI Europa ICTS and the Telecom Industry Representatives: Attacks Against Telecom Employees Must Stop Now*. www.gsma.com/gsmaeurope/news/joint-statement-of-the-uni-europa-icts-and-the-telecom-industry-representatives-attacks-against-telecom-employees-must-stop-now/.

21. Telecommunications Forum, 15 May 2020. www.tcf.org.nz/con sumers/news/2020-05-15-mobile-operators-warn-arson-attempts-on-cell-sites-may-impact-phone-and-internet-connectivity/.
22. C. Keall, 14 April 2020. Covid 19 Coronavirus: Facebook Responds after NZ Cell Tower Arson Brag Video Goes Viral on its Platform. *NZ Herald.* www.nzherald.co.nz/business/covid-19-coronavirus-facebook-responds-after-nz-cell-tower-arson-brag-video-goes-viral-on-its-platfo rm/5HJYAL3YTW5AYDYOGEF7YG3YMA/.
23. J. Weekes, 20 May 2020. Attacks on 5G Cellphone Towers Show 'Idiotic' Social Media Conspiracy Theories Growing Threat to NZ, Analyst Says. *Stuff.* www.stuff.co.nz/national/crime/300015974/attacks-on-5g-cellphone-towers-show-idiotic-social-media-conspiracy-theories-growing-threat-to-nz-analyst-says.
24. V. Molyneux, 4 April 2020. Prime Minister Jacinda Ardern Says Conspiracy Theories Linking 5G and COVID-19 Are 'Just Not True'. *Stuff.* www.newshub.co.nz/home/politics/2020/04/prime-minister-jacinda-ardern-says-conspiracy-theories-linking-5g-and-covid-19-are-just-not-true.html.
25. C. Keall, 9 June 2020. Vodafone, Spark Take Security Steps after Three More '5G' Celltower Fires – Taking Total to 18. *NZ Herald.* www.nzherald.co.nz/business/news/article.cfm?c_id=3&objectid=12338313.
26. M. Daalder, 26 May 2020. The Pandemic Is Facebook's Ultimate Test – and It's Failing. *Newsroom.* www.newsroom.co.nz/2020/05/26/1181114/the-pandemic-is-facebooks-ultimate-test-and-its-failing.
27. Telecommunications Forum, 15 May 2020.
28. J. Huakau, July 2014. *Locality Population Snapshot South Auckland.* Waipareira Tuararo, Te Whānau o Waipareira, Research Unit. www.waipareira.com/wp-content/uploads/2017/11/TPM6.-Locality-Population-Snapshot-SOUTH-Auckland.pdf.
29. K. Taiuru, 29 May 2020. www.sciencemediacentre.co.nz/2020/05/29/understanding-5g-concerns-expert-qa/.
30. D. O'Carroll, 2013. An Analysis of How Rangatahi Maori Use Social Networking Sites. *Mai Journal*, 2(1). www.journal.mai.ac.nz/sites/default/files/Vol%202%20%281%29%20OCarroll.pdf.
31. B. Masters-Awatere and L.W. Nikora, 2017. *Indigenous Programmes and Evaluation: An Excluded Worldview.* New Zealand Council for Educational Research. www.nzcer.org.nz/nzcerpress/evaluation-matters.
32. J. Berentson-Shaw, 2011. *A Matter of Fact: Talking Truth in a Post-truth World.* https://books.google.co.nz/books?id=WCppDwAAQBA J&pg=PA6&lpg=PA6&dq=maori+misinformation+more&source=bl &ots=MZ5iOqBEMK&sig=ACfU3U0MFuCJJa_sRWQFOUOK-Cqc4O1mjg&hl=en&sa=X&ved=2ahUKEwiRw7LJh-_pAhVnwTg GHQuyC2g4FBDoATAEegQIChAB#v=onepage&q=maori%20 misinformation%20more&f=false.
33. T.P. Kōkiri, January 2010. *Use of Broadcasting and e-Media, Māori Language and Culture.* www.tpk.govt.nz/en/a-matou-mohiotanga/

broadcasting/use-of-broadcasting-and-e-media-maori-language-and/
online/9.

34. E. Pacheco and N. Melhuish, 2019. Exploring New Zealand Children's
Internet Access, Skills and Opportunities. *Netsafe*. www.netsafe.org.
nz/wp-content/uploads/2019/09/NZ-childrens-technology-access-use-
skills-opportunities-2019-3.pdf.

35. D. Cormack, R. Harris, and J. Stanley, 2013. *Māori Experiences of Multiple Forms of Discrimination: Findings from Te Kupenga*. www.tandfon
line.com/doi/full/10.1080/1177083X.2019.1657472.

36. K. Williams, 30 November 2020. Our Truth, Tā Mātou Pono: Stuff
Introduces New Treaty of Waitangi Based Charter Following Historic
Apology. *Stuff*. www.stuff.co.nz/pou-tiaki/our-truth/123533668/our-
truth-t-mtou-pono-stuff-introduces-new-treaty-of-waitangi-based-
charter-following-historic-apology.

37. Human Rights Act, 1993. www.legislation.govt.nz/act/public/1993/
0082/latest/DLM304212.html.

38. S. Simmonds, 2019. *Evidence Review of Effective Risk Reduction Inter-
ventions for Māori whānau and Communities*. https://fireandemergency.
nz/assets/Documents/Files/Report_170_Maori_interventions_evi
dence-_review.pdf.

39. H. Gifford and S. Bradbrook, 2009. *Recent Actions by Māori Politicians
and Health Advocates for a Tobacco-Free Aotearoa/New Zealand, A Brief
Review*. Wellington: Te Reo Mārama. www.otago.ac.nz/wellington/
otago022879.pdf.

40. S. Elders, F. Nelson, and R. Johnson, 2015. *Māori Perspectives of Public
Information Advertising Campaigns*. Auckland University of Technol-
ogy. https://www-s3-live.kent.edu/s3fs-root/s3fs-public/file/03-Steve-
Elers-Frances-Nelson-Rosser-Johnson.pdf.

41. Y.X. Poon, 11 September 2020. How Taiwan Used Memes to Fight
Pandemic Rumours. *GovInsider*. https://govinsider.asia/inclusive-gov/
audrey-tang-digital-minister-how-taiwan-used-memes-to-fight-pan
demic-rumours/.

42. Vodafone, 10 December 2019. *New Zealanders and New Zealand Businesses Can Now Join in the Global 5G Mobile Party*. https://news.vodafone.
co.nz/business/new-zealanders-and-new-zealand-businesses-can-now-
join-in-the-global-5g-mobile-party.

43. TCF, 15 May 2020. *Mobile Operators Warn Arson Attempts on Cell
Sites May Impact Phone and Internet Connectivity*. www.tcf.org.nz/con
sumers/news/2020-05-15-mobile-operators-warn-arson-attempts-on-
cell-sites-may-impact-phone-and-internet-connectivity/.

44. TCF, 16 September 2020. *Telecommunications Industry Launches New
5G Facts Website*. www.tcf.org.nz/consumers/news/2020-09-16-telecom
munications-industry-launches-new-5g-facts-website/.

45. Vodafone New Zealand, 14 July 2020. *Vodafone Aotearoa Unveils 'Hon-
ouring the Principles of the Treaty of Waitangi' Policy as Part of Māori
Development Strategy*. https://news.vodafone.co.nz/treaty-policy.

46. New Zealand Ministry for Culture and Heritage, 17 May 2017. *The Treaty
in Brief*. https://nzhistory.govt.nz/politics/treaty/the-treaty-in-brief.

12

THE TRAGEDY OF THE "TRAGEDY OF THE COMMONS" METAPHOR

MARTIN SIMS

Contents

Metaphors are much more important than they seem. Saying something has the characteristics of something else is a very common figure of speech, and it is tempting to think that it is just an occasional aid to understanding.

For the philosopher Nietzsche, this was a fundamental misapprehension. He argued that what we regard as "truth" is actually a "mobile army of metaphors, metonyms, and anthropomorphisms" but because we use them so often they become engrained into our culture and personal understanding. "Truths are illusions about which one has forgotten that this is what they are; metaphors which are worn out and without sensuous power; coins which have lost their pictures and now matter only as metal, no longer as coins."[1]

This perspective can also be applied to how we think about the use of spectrum. de Vries argues that "relatively abstract concepts like spectrum and signal are partially understood in terms of the more concrete concepts which are directly grounded in our experience, like space, objects, sound, and people."[2] These mental models "shape our thinking in ways that are not directly coupled to the technical details

DOI: 10.1201/9781003156765-13

of communications." Spectrum is often seen as land, which de Vries calls a "rhetorical trap": "since most land is privately held; public spaces are the exceptions that prove the rule, and most people have no experience at all of commons land."[3] Other models are possible – we could, for example, think of spectrum as an ocean, with its suggestions of boundless availability. All of these models "highlight some attributes and hide others."[4]

None of this means that the use of metaphors is wrong or avoidable – they not only enable our thinking about a subject but also limit it. "Battles of ideas are fought over which metaphor is appropriate," argues the psychologist Professor Nick Chater. "Is light made up of particles or waves? . . . Is nature a harmonious society or a brutal war of all against all? Such metaphors are not marginal to thought but its very essence."[5]

Metaphorical Muscle

The power of metaphors in framing debate was illustrated in a recent study, based on just the use of a single word. In the USA, Thibodeau and Boroditsky[6] showed two sets of participants two brief articles about crime in a fictional city. There were only a few words difference between each article: for one group, crime was described as a "beast preying" on the city. For the other half, crime was described as a "virus infecting" the city.

The virus group "universally suggested investigating the source of the virus and implementing social reforms and prevention measures to decrease the spread of the virus." The beast group "universally suggested capturing the beast and then killing or caging it. They wanted to organize a hunting party or hire animal control specialists to track down the beast and stop it from ravaging the city."[7] The differences are quite profound: the first group leans towards prevention and rehabilitation, while the second group favours punishment.

This single metaphor induced significant difference in support for social policies, going far beyond the participant's pre-existing political affiliation. "Metaphors in language appear to instantiate frame-consistent knowledge structures and invite structurally consistent inferences. Far from being mere rhetorical flourishes, metaphors have profound influences on how we conceptualize and act with respect to important societal issues,"[8] the authors say. They argue that metaphors are particularly important in "discussions of social policy, where it often seems impossible to 'literally' discuss immigration, the economy, or crime."[9]

So, what are the implications for spectrum, where policy decisions are connected not only to social policy but also to technology and economic policy? The power of metaphor and framing are equally important, especially considering that most of the politicians, who make the ultimate decisions, are not spectrum experts.

A Popular Misconception

This chapter focuses on a metaphor that has been very influential in spectrum policy: the tragedy of the commons. Its power seems particularly unfair because there is so little evidence to support it.

The term is often used as if it refers to a "known" which emerged from the study of land usage: the idea that commonly held resources go to rack and ruin. In fact, the opposite is the case: in the disciplines where it has been studied, it is widely viewed as discredited. In history, there is little evidence of a tragedy, and in economics the academic who challenged the concept won a Nobel prize.

Even the person who coined the term recanted. Modern usage stems from a 1968 academic article called *The Tragedy of the Commons* by Garrett Hardin.[10] He argued that open pastures are inevitably overgrazed because adding extra cattle gives a large individual benefit to their owner, whereas the eventual environmental degradation caused by overstocking is shared by all users. Privatise common land and the problem disappears: overstocking damages the landowner directly, so this incentive to inefficiency disappears.

In the spectrum world, this argument claims that congestion will eventually make unlicensed bands unusable because individuals suffer only marginally as the problem intensifies.

Hardin was extending an argument made by the Victorian economist William Forster Lloyd in 1833 and used to justify enclosure of common lands in the UK. Enclosure was the practice of taking lands used by the public at large and giving or selling them to private landlords. This had been going on since medieval times but intensified from the 17th century onwards.

So where is the evidence that over-grazing caused by common ownership leads to Lloyd's famously "puny and stunted" cattle? In fact, there isn't much. To paraphrase the historian E.P. Thompson, the problem with the "tragedy" argument is that it assumes commoners have no common sense. They were are aware of the problems of over-grazing and created rules and procedures to prevent this. Scholars like Neeson[11] have documented how communities held special courts to manage the commons.

Furthermore, who would have the money to buy the additional animals to ruin the grazing for everyone else? The richer farmers who would have the most to gain from enclosing common land! Neeson presents evidence of the well-off being prosecuted for ignoring the rules of the common.[12]

Looking at English history, is the "tragedy of the commons" a myth? There seem to be isolated examples of badly managed commons but plenty of examples of good management.

Hardin himself recognised this and retracted his original thesis, conceding: "The title of my 1968 paper should have been 'The Tragedy of the Unmanaged Commons.'"[13]

The economist Elinor Ostrom won the Nobel prize in 2009 for her re-examination of this debate, producing more subtle and convincing arguments based on the study of the management of natural local resources in developing countries. She demonstrated that there were a wide range of successful methods used to manage common resources. For her, it was the involvement of the local community that was the key to success.

In Whose Interests?

By providing the justification for enclosure, the "tragedy of the commons" argument benefitted the higher echelons of society who would come to own the former common lands and farm them for private profit. Its claims of inevitability and increased efficiency did not convince the users of common land who realised their economic interests would suffer. From the 14th to 17th centuries commoners opposed enclosure by petitioning the king, through legal action, by riots and even rebellions.[14]

An interesting variation of the anti-commons argument was used to justify the draining of wide areas of marshland (usually called fens) in Eastern England from the 17th to 19th centuries, turning it into private farming land. The fenland included fields that flooded in the winter, providing fertile arable land in the summer and wide areas of marsh where local people could catch fish and wildfowl. They were managed through very detailed rules, the best known being the 72 articles of the 1548 "Fen Charter" which included measures such as a close season on wildfowling and what we might describe today as fishing quotas.[15]

In medieval times, the fens were some of England's richest areas: their tax assessment was the fourth highest in the country.[16] Wealth was also more evenly distributed than in other areas because the

Figure 12.1   The kitchen in a student house: a poorly managed commons.

Source: Wikimedia.

abundance of common land meant that even the landless had some-where to graze cattle.[17]

However, in 1620, King James, keen to swell his coffers by the acquisition of new private farming land, disingenuously claimed that for the "honour of his kingdom" he could not allow the fens to "lie waste and unprofitable."[18] A few years earlier a drainage proposal claimed the fens were a "vague deserted Empire without population" which would be turned into a fertile region, with "wild and useless products" becoming an "abundance of grain and pasture."[19]

These were self-interested arguments used to justify a land-grab with no more substance than the "tragedy of commons" rationale for enclosure. Like enclosure, drainage was fiercely resisted by people liv-ing in the fens, as documented by James Boyce in his book *Imperial Mud*.[20]

A Metaphorical New Lease of Life

From the 1970s onwards, the "tragedy of the commons" metaphor was enthusiastically adopted by free marketeers as a justification for the privatisation of what were previously common resources, such as forests, fishing and agricultural land, often in the developing world.[21]

The inevitability aspect of Harbin's argument forms a large part of the appeal. It presumes that human beings are *homo economicus:* ratio-nal agents motivated by narrow self-interest. Human nature, there-fore, means we are predetermined to destroy any common resource. However, historical studies and Ostrom's work show users were able to take a long-term view and accept communal rules designed to pre-vent over-use.

A recent example of the "inevitability" argument was made in the context of the debates about a variant of the LTE mobile standard (LTE-U) having access to the licence exempt bands. Tim Worstall of the free-market Adam Smith Institute argued that the arrival of this new technology meant that governments would need to licence and sell bands that were previously unlicensed:

> Simply because as demand rises on this limited resource without such
> restrictions of access then no one will be able to use it as too many

signals interfere with each other. [. . .] And there's obviously a certain wryness to the point. It's exactly those currently rushing to exploit a currently free resource which means it cannot be free in the future.[22]

This is an example of a metaphor limiting our understanding of a subject. Worstall's unquestioning acceptance of the tragedy of the commons argument leads him to ignore the success of mutually agreed rules in managing a commons, to assume that the same principles are applicable to all shared resources and to ignore the specifics of markets and technologies.

The equipment standards requirements required in the unlicensed bands are the equivalent of the rules developed to prevent over-grazing in commons. They have been mutually agreed in standards bodies and are very successful in preventing interference. A study by the UK regulator, Ofcom,[23] found that any congestion in the 2.4 GHz band was due not to a proliferation of Wi-Fi devices but was instead due to other devices using the same spectrum, particularly analogue video senders. In other words, devices that were not obeying the mutually agreed rules of the common.

And do these shared resources have similar characteristics? How analogous are pastures and unlicensed spectrum? How similar are cows and Wi-Fi? If one cow eats a square metre of grass, another cow cannot eat that same piece of grass, at least not until it regrows a few weeks later. But unlike grass, unlicensed spectrum is re-usable because of the low power limits required. Even in the most congested situations, like blocks of flats close to each other, most Wi-Fi signals will stop at the dividing walls. Where they do not, Wi-Fis can use vacant channels within the unlicensed bands. This means that the same spectrum can be re-used within 100 metres or less.

Even if we agree that every individual is incentivised to buy more cows to maximise their income from the shared pasture, does the same incentive exist for devices operating in wireless spectrum? Certainly not for individuals – why would anybody want more than one Wi-Fi if they are already getting good wireless coverage? Arguably manufacturers have an incentive to build IoT connectivity into more and more devices and most of these use the wireless bands. But there is a limit to the number of devices in each home which can benefit from connectivity. It is not the infinite number of cows which *homo economicus* would seek to release on the shared pasture to pursue profit maximisation.

Furthermore, Worstall's belief in inevitability leads him to make unworkable proposals. Buying the currently unlicensed bands doesn't seem to be in the interests of any market players. Why would mobile operators want to buy something – Wi-Fi off-load – that they currently get for free and prevent data overload on their networks?[24] Why would fixed line players – who depend on Wi-Fi to deliver broadband connectivity to the consumer – want to burden themselves with extra costs? Far better for market players to support the identification of further unlicensed spectrum if the existing bands are insufficient.[25]

Congestion in the unlicensed bands is certainly not impossible, particularly as consumers demand faster download speeds. It has become a concern for spectrum managers worldwide, and many countries are making 6 GHz an unlicensed band to complement 2.4 GHz and 5.8 GHz. We should also note that as more Wi-Fi devices have appeared the technological standards have improved to reduce interference. To return to the commons metaphor, it is like having a pasture where the speed at which grass can grow is ever-increasing.

What Use Is a Tragedy?

What can spectrum policy learn from work in history and economics about the management of common resources? A simple summary is that there are numerous examples of common resources being managed effectively for wider social benefit. There are examples of the over-use of commons to the public detriment – the "tragedy of the commons" – but these are rare and have been used as a justification for a practice of enclosure which overwhelmingly disadvantaged the poor.

Debate about spectrum policy has been disproportionately influenced by the "tragedy of the commons," one of the weaker claims in a much wider debate about the economics of common resources. We should certainly not regard the "tragedy of the commons" as an inevitability – the evidence does not support this.

Perhaps Ostrom's law might be a more useful mental framework. Based on the economist's study of successful commons management, this states that an allocation of resources that works in practice can work in theory.[26] In practice, spectrum management is a mixed economy using a range of property regimes, including licensed, unlicensed, leasing and sharing.

The unlicensed model has been a notable success and now supports billions of devices. It is not doomed to failure as the "tragedy of the commons" argument suggests. In fact, it is so successful that there is considerable backing for expanding this approach into new bands, such as 6 GHz.[27] The mobile sector generally supports a smaller extension, with the upper part of the band given over to licensed use.[28]

The success of the mobile industry was founded on licensed spectrum, and one might expect them to oppose to the unlicensed approach. In fact, they are enthusiastic users and support the expansion of unlicensed bands, albeit not as large an expansion as other stakeholders whose reliance on Wi-Fi is greater. The mobile industry doesn't believe the commons approach is an impending tragedy which is doomed to failure – they see unlicensed spectrum as a successful means of resource management which should be extended further.

This is what happens in practice and it doesn't fit the "tragedy of the commons" metaphor. It is more like the complex web of negotiation and renegotiation between the public and the private designed to accommodate a wide variety of needs as characterised in Ostrom's work. The "tragedy of the commons" is of very limited application in spectrum policy and generally does more to obscure than enhance our understanding.

Notes

1. F. Nietzsche, 1873. On Truth and Lies in a Nonmoral Sense.
2. Imagining Radio: Mental Models of Wireless Communication, 2007. 2nd IEEE International Symposium on New Frontiers in Dynamic Spectrum Access Networks, pp. 372–380. doi: 10.1109/DYSPAN.2007.55.
3. Ibid.
4. Ibid.
5. N. Chater, 2018. *The Mind Is Flat*, p. 210. UK: Allen Lane.
6. P.H. Thibodeau and L. Boroditsky, 2011. Metaphors We Think with: The Role of Metaphor in Reasoning. *PLOS ONE*, 6(2), pp. e16782. https://doi.org/10.1371/journal.pone.0016782.
7. Ibid.
8. Ibid.

9. Ibid.

10. G. Hardin, 13 December 1968. The Tragedy of the Commons. *Science, New Series,* 162(3859), pp. 1243–1248 https://pages.mtu.edu/~asmayer/rural_sustain/governance/Hardin%201968.pdf.

11. J.M. Neeson, 1996. *Commoners: Common Right, Enclosure and Social Change in England, 1700–1820,* pp. 153–157. Toronto: York University.

12. Ibid., p. 152.

13. G. Hardin, 1994. *The Tragedy of the Unmanaged Commons.* USA: Elsevier

14. See S. Fairlie, 2009. A Short History of Enclosure in Britain. *The Land Magazine,* p. 20. UK. He argues enclosure was a factor in the 1381 Peasants' Revolt, Jack Cade's rebellion of 1450 Kett's rebellion of 1549 and Captain Pouch revolts of 1604–1607.

15. J. Boyce, 2020. *Imperial Mud,* p. 33. UK: Icon Books

16. Ibid., p. 21.

17. Ibid., p. 23.

18. Ibid., p. 39.

19. Ibid., p. 35 – the proposal was made in 1589.

20. Ibid.

21. S. Partelow, D.J. Abson, A. Schlüter, M. Fernández-Giménez, H. von Wehrden, and N. Collier, 2019. Privatizing the Commons: New Approaches Need Broader Evaluative Criteria for Sustainability. *International Journal of the Commons,* 13(1), pp. 747–776, p. 3. http://doi.org/10.18352/ijc.938.

22. T. Worstall, 24 February 2016. Free Cellphone Bandwidth and the Tragedy of the Commons. *Forbes Magazine.* www.forbes.com/sites/timworstall/2016/02/24/free-cellphone-bandwidth-and-the-tragedy-of-the-commons/.

23. Mass Consultants Ltd, 2009. *Estimating the Utilisation of Key License-Exempt Spectrum Bands.* UK: Ofcom

24. In 2017 54 percent of total mobile data traffic was offloaded onto Wi-Fi, according to Cisco's Mobile Visual Networking Index. https://newsroom.cisco.com/press-release-content?type=webcontent&articleId=1967403.

25. For reasons of space I have not addressed the consumer implications of selling unlicensed bands, which include likely added cost, difficulties in maintaining competition the practicalities of roaming between networks.

26. For a more detailed discussion see L.A. Fennell, 1 February 2011. Ostrom's Law: Property Rights in the Commons. *International Journal of the Commons,* 5(1), pp. 9–27. University of Chicago Institute for Law & Economics Olin Research Paper No. 584, SSRN: https://ssrn.com/abstract=1962336.

27. At the time of writing this had been agreed in the USA and Brazil, with European countries supporting unlicensed use in the lower part of 6 GHz.

28. See 6 GHz Overview, *Policy Tracker,* 16 January 2020. www.policytracker.com/bands/6-ghz-2020/.

Index